# Mathematics

**FOR COMMON ENTRANCE**

**13+**

## Exam Practice Questions

# Mathematics

## FOR COMMON ENTRANCE

13+

## Exam Practice Questions

David E Hanson

GALORE PARK

AN HACHETTE UK COMPANY

# About the author

David Hanson has over 40 years' experience of teaching. For a number of years he was Leader of the ISEB 11+ Maths setting team, a member of the ISEB 13+ Maths setting team and a member of the ISEB Editorial Endorsement Committee. He edited the *SATIPS Maths Broadsheet* for many years. David has retired from teaching to run a small shop trading in collectors' items.

# Acknowledgements

I would like to thank Gina de Cova for her assistance at various early stages of this book.

David E Hanson
August 2015

Every effort has been made to trace all copyright holders, but if any have been inadvertently overlooked, the Publishers will be pleased to make the necessary arrangements at the first opportunity.

Although every effort has been made to ensure that website addresses are correct at time of going to press, Galore Park cannot be held responsible for the content of any website mentioned in this book. It is sometimes possible to find a relocated web page by typing in the address of the home page for a website in the URL window of your browser.

Hachette UK's policy is to use papers that are natural, renewable and recyclable products and made from wood grown in sustainable forests. The logging and manufacturing processes are expected to conform to the environmental regulations of the country of origin.

Orders: please contact Bookpoint Ltd, 130 Milton Park, Abingdon, Oxon OX14 4SB. Telephone: (44) 01235 827720. Fax: (44) 01235 400454. Email education@bookpoint.co.uk Lines are open from 9 a.m. to 5 p.m., Monday to Saturday, with a 24-hour message answering service. Visit our website at www.galorepark.co.uk for details of other revision guides for Common Entrance, examination papers and Galore Park publications.

**ISBN: 978 1 471846 92 2**

© David E Hanson 2015

First published in 2015 by
Galore Park Publishing Ltd,
An Hachette UK Company
Carmelite House
50 Victoria Embankment
London EC4Y 0DZ
www.galorepark.co.uk

Impression number 10  9  8  7
Year 2019

Some illustrations by Ian Moores were re-used. All other illustrations by Aptara, Inc.
Typeset in India
Printed in Great Britain by CPI Group (UK) Ltd, Croydon CR0 4YY

A catalogue record for this title is available from the British Library.

# Contents

Introduction   vi

1   Number   1

    1.1   Properties of numbers   1

    1.2   Fractions, decimals, percentages; ratio   4

2   Calculations   11

    2.1   Mental strategies   11

    2.2   Written methods   18

    2.3   Calculator methods   19

3   Problem solving   22

    3.1   Reasoning about numbers or shapes   22

    3.2   Real-life mathematics   33

4   Algebra   43

    4.1   Equations and formulae   43

    4.2   Sequences and functions   52

5   Geometry and measures   66

    5.1   Measures   66

    5.2   Shape   75

    5.3   Space   79

6   Statistics and probability   100

    6.1   Statistics   100

    6.2   Probability   114

# Introduction

## → The curriculum and the examination syllabus

The mathematics curriculum and the examination syllabus are subject to relatively minor changes or emphases from time to time, whereas the body of mathematical skills and knowledge which teachers consider valuable seems to remain fairly constant.

For completeness, and to allow greater flexibility in the use of this material, some questions included here may be outside the syllabus currently examined, even though they are likely to be within the capability of the majority of students in most schools. It is left to teachers to select questions which they consider appropriate and it is assumed that teachers will wish to differentiate according to student abilities.

The material is appropriate for KS3 studies but, for completeness, questions cover ideas met in all years up to Year 8.

The contents pages outline the way in which questions have been grouped in 6 'strands': Number; Calculations; Problem solving; Algebra; Geometry and measures; Statistics and probability. The sections within the strands have been numbered for easier reference.

### Examination levels

Level 1 and Level 2 papers are based upon the same syllabus, but *some* of the questions in Level 1 papers will be generally more accessible.

Questions that are more demanding and *may* be more appropriate for candidates taking papers at Level 2 and above are indicated by this Level 2 symbol in the margin:

Many of these questions will be accessible to all students and it is assumed that teachers and parents will wish to encourage students to attempt questions that may be just beyond the requirements for the intended examination.

It is expected that the majority of students will take Level 2 papers.

Level 3 and CASE papers are based upon the extended syllabus.

Questions appropriate for Level 3 and CASE candidates, often involving the extended syllabus, are indicated by this Level 3 symbol in the margin:

Some extension questions are included for the interest of capable students and are indicated in the text.

*The level notes are included as a guide only and most of the questions are suitable for students taking papers at any level.*

*It is strongly recommended that reference is made regularly to the current Examination Syllabus and to recent past papers.*

The following table shows how the contents of this book relate to the ISEB syllabus.

| Chapter | ISEB syllabus |
|---|---|
| **1 Number** | |
| 1.1 Properties of numbers | Number |
| 1.2 Fractions, decimals, percentages; ratio | Ratio, proportion and rates |
| **2 Calculations** | |
| 2.1 Mental strategies | Number |
| 2.2 Written methods | |
| 2.3 Calculator methods | |
| **3 Problem solving** | |
| 3.1 Reasoning about numbers or shapes | *Reason mathematically* *Solve problems* |
| 3.2 Real-life mathematics | *Develop fluency* *Solve problems* |
| **4 Algebra** | |
| 4.1 Equations and formulae | Algebra |
| 4.2 Sequences and functions | |
| **5 Geometry and measures** | |
| 5.1 Measures | Geometry and measures |
| 5.2 Shape | |
| 5.3 Space | |
| **6 Statistics and probability** | |
| 6.1 Statistics | Statistics |
| 6.2 Probability | Probability |

# → Using this book

The book has been designed for use by students, under the guidance of a teacher or parent, as a resource for practice of basic skills and recall of knowledge.

It is assumed that, in addition to plain paper, the following grids will be available for use where appropriate:

- centimetre squared

- centimetre isometric dotted

- centimetre square dotted

- graph (cm and 2 mm)

**If students are permitted to draw in this book, then valuable time may be saved.**
Students are expected

- to show full working where appropriate and, at all times, to make their method clear to the marker

- to produce a personal record of achievement which will prove valuable as an additional revision aid.

It is assumed that, throughout, students will

● make use of estimation skills

● pay attention to the order of operations (BIDMAS or BODMAS)

● use strategies to check the reasonableness of results

● use a calculator *only* when instructed or allowed to do so.

Whilst this book has been compiled for use in independent schools, it is expected that it will also prove useful for students in state schools and home schoolers. Answers can be checked in the separate *Answer Book*.

## → The questions

Questions follow the ISEB format and are numbered either:

1 (a)                    (b)                    (c) where parts of questions are *not* related.

or

1 (i)                    (ii)                    (iii) where parts of questions *are* related.

Almost all of the questions are modelled on questions from past Common Entrance 13+ papers, using similar wording and mark allocation.

Within each broad group of questions, some grading in difficulty has been attempted and harder questions may be found towards the end of each grouping. Many of these harder questions will be within the capabilities of most students.

Many questions involve several skills. These questions have not been split but have been placed wherever seemed most appropriate.

The number of questions on a particular topic reflects the frequency with which such questions have appeared in the Common Entrance papers.

## → Calculators

Questions in 1.1, 1.2, 2.1 and 2.2 should be tackled *without* a calculator.

In 3.1, 4.2, 5.2 and 6.1 a calculator should not be needed.

Questions in 2.3 *require* the use of a suitable calculator.

Questions which involve both calculator and non-calculator parts have the parts clearly indicated.

It is assumed that students will

● be encouraged to tackle all other questions *without* the use of a calculator

● have the opportunity to decide for themselves when the use of a calculator is appropriate and when other methods are more effective.

## → Tips on taking the exam

### Before the exam

● Have all your equipment ready the night before. You will need: calculator, pens, pencils, rubber, pencil sharpener, ruler, protractor, compasses, set square.

● Make sure you are at your best by getting a good night's sleep before the exam.

● Have a good breakfast in the morning.

● Take some water into the exam if you are allowed.

● Think positively and keep calm.

# 1 Number

## 1.1 Properties of numbers

In this section the questions cover the following topics:

- Multiples and factors
- Prime numbers
- Negative numbers
- Place value
- Ordering
- Estimation and approximation
- Numbers written in Roman numerals

Level 3

- Standard form

Many questions cover several topics.

*Questions must be answered without using a calculator.*

1   From the following list

| 30 | 31 | 32 | 33 | 34 | 35 | 36 | 37 | 38 | 39 | 40 |

choose *one* number which is

(i)    a prime number                    (1)         (iii)  a square number              (1)

(ii)   a multiple of 5 and 7           (1)         (iv)  a factor of 1000            (1)

2   From the following list of numbers

| 3 | 5 | 13 | 21 | 36 | 64 |

write down

(i)    a multiple of 18                   (1)         (iv)  the square root of 25      (1)

(ii)   a factor of 18                       (1)         (v)   a cube number.              (1)

(iii)  the product of 3 and 7        (1)

3  From this list of numbers

| 1 | 2 | 5 | 9 | 12 | 15 | 18 | 25 | 27 | 28 |

write down any two numbers which are

(i)   multiples of 4                    (1)        (iv)  square numbers              (2)

(ii)  factors of 30                     (1)        (v)   cube numbers.              (2)

(iii) prime                             (2)

4  (a)  Find the value of $2 + 3 \times 4$                                          (2)

   (b)  Find the value of $5^2 - 4^3$                                               (3)

   (c)  Write down two prime numbers whose sum is 30                                (2)

   (d)  Write 150 as the product of prime factors, using indices.                   (3)

5  (a)  Express the number 560 as the product of prime factors, using indices.  (3)

   (b)  Evaluate $2^2 \times 3^3 \times 5^2 \times 7$                                (3)

6  (i)   Express the number 56 as the product of prime factors, using indices.  (2)

   (ii)  Find the smallest number that can be multiplied by 56 to give a
         perfect square.                                                         (1)

7  (i)   Express the number 495 as the product of prime factors, using indices. (3)

   (ii)  Use your answer to part (i) to find a common factor of 66 and 495     (1)

8  (i)   Express the number 54 as the product of prime factors, using indices.  (2)

   (ii)  $24 = 2^3 \times 3$

         What is the largest number which will divide exactly into 24 and 54?   (1)

9  (i)   Express the number $24 \times 32$ as the product of prime factors,
         using indices.                                                          (3)

   (ii)  What is the lowest integer that $24 \times 32$ can be multiplied by to
         produce a perfect square?                                              (1)

10 Copy and complete the following:

   (i)   (a)  $^-3 + 5 =$            (1)        (c)  $7 - \,^-4 =$                (1)

         (b)  $4 - 7 =$             (1)        (d)  $5 + \,^-1 - \,^-2 =$        (1)

(ii) (a) $5 \times {}^-2 =$ (1)  (c) $({}^-4)^2 =$ (1)

(b) ${}^-3 \times {}^-2 =$ (1)  (d) ${}^-4 \div 2 =$ (1)

11 (a) (i) What is the value of the 7 in 37 405? (1)

(ii) What is the value of the 3 in 1.035? (1)

(b) Write the following in order of *increasing* size: (3)

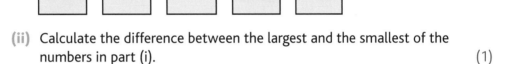

9.05 | 0.59 | 9.5 | 5.9 | 0.95 | 5.09

(c) Write the following in order of *decreasing* size: (3)

12.34 | 13.42 | 14.32 | 12.43 | 13.24 | 14.23

12 (i) Write the following numbers in order of size, starting with the smallest: (3)

5.505 | 5.5 | 5 | 5.055 | 5.55

(ii) Calculate the difference between the largest and the smallest of the numbers in part (i). (1)

13 Arrange the following numbers in order of size, starting with the smallest: (3)

6.4 | $6\frac{1}{3}$ | 6.06 | 6.006 | 6.51

14 Arrange these number cards in ascending order of size, starting with the smallest: (3)

66% | $\frac{13}{20}$ | 0.67 | $\frac{2}{3}$

15 (a) Round:

(i) 3406 to the nearest ten (1)

(ii) 150 to the nearest hundred (1)

(iii) 59.45 to the nearest whole number. (1)

(b) Write:

(i) 27.56 to 1 decimal place (1)

(ii) 0.344 to 2 decimal places (1)

(iii) 35.19 to 3 significant figures (1)

(iv) 0.3095 to 3 significant figures. (2)

16 (a) Estimate, to the nearest whole number:

    (i) $5.9 \times 6.1$     (1)     (ii) $23.8 \div 7.8$     (2)

  (b) Estimate, to 1 significant figure:

    (i) $11.9 \times 4.9$     (1)     (ii) $1399 \div 34$     (3)

17 (a) Estimate the value of $\dfrac{29.7 \times 0.62}{2.95}$

    giving your answer to 1 significant figure.     (2)

  (b) A circle has radius 29 cm.

    Estimate the area of the circle (in cm²), giving your answer to 1 significant figure.     (3)

18 (a) Write the following as Roman numbers:

    (i) 25     (1)     (iii) 1066     (1)

    (ii) 107     (1)     (iv) 2015     (1)

  (b) Write the following Roman numbers as ordinary numbers:

    (i) XXVIII     (1)     (iii) DCCLXXXVII     (1)

    (ii) LVI     (1)     (iv) MMDCLXVI     (1)

**Level 3**

19 (a) Write the following numbers in standard form:

    (i) 4900     (1)     (ii) 0.000 75     (1)

  (b) Write the following as ordinary numbers:

    (i) $1.8 \times 10^6$     (1)     (ii) $3.2 \times 10^{-3}$     (1)

# 1.2 Fractions, decimals, percentages; ratio

In this section the questions cover the following topics:

- Fractions
- Decimals
- Percentages
- Ratio and proportion

Many questions cover several topics.

*In this section the questions must be answered without using a calculator.*

1 (a) (i) Write the fraction representing 3 parts of a whole containing 7 parts.     (1)

(ii) What fraction of this rectangle has been shaded? (2)

(b) Change the fraction $\frac{2}{5}$ to:

   (i)   an equivalent fraction with denominator 20 (1)

   (ii)  a decimal (1)

   (iii) a percentage. (1)

2  (a) Write the fraction $\frac{15}{24}$ in its lowest terms (simplest form). (1)

   (b) (i)  Write the improper fraction $\frac{11}{3}$ as a mixed number. (1)

      (ii) Write the mixed number $4\frac{1}{2}$ as an improper fraction. (1)

   (c) (i)  Write the mixed number $3\frac{1}{4}$ as a decimal. (2)

      (ii) Write the fraction $\frac{7}{20}$ as a percentage. (2)

3  Copy and complete the following table showing rows of equivalent
   fractions, decimals and percentages: (5)

| Fraction | Decimal | Percentage |
|----------|---------|------------|
| $\frac{7}{20}$ | 0.35 | 35% |
| $\frac{1}{4}$ | | |
| | 1.2 | 120% |
| | | 15% |

4  (a) Express 26% as a fraction in its lowest terms. (2)

   (b) Write $\frac{13}{20}$ as a decimal. (2)

   (c) Calculate $\frac{2}{3}$ of £4.80 (2)

5  (i) (a) Write 45% as a decimal. (2)

      (b) Write $\frac{2}{5}$ as a decimal. (2)

      (c) Write $\frac{4}{9}$ as a decimal correct to 2 decimal places. (2)

   (ii) Use your answers to part (i) to write the following numbers in
        ascending order, starting with the smallest: (1)

   45%     $\frac{2}{5}$     $\frac{4}{9}$

6  (a)  Express $\frac{5}{8}$ as a decimal correct to 2 decimal places.                                    (2)

   (b)  Express 7 as a percentage of 20                                                                     (2)

   (c)  Express 0.18 as a fraction in its lowest terms.                                                     (2)

   (d)  Find 8% of £260                                                                                     (2)

7  (a)  Find the value of $\frac{2}{3}$ of £45.60                                                           (2)

   (b)  (i)   Write 80% as a fraction in its lowest terms.                                                  (2)

        (ii)  Find 80% of £450                                                                              (2)

   (c)  Calculate 97 cm + 2.05 m. Give your answer in metres.                                               (2)

8  (a)  Mr Smith earns £2200 a month. He spends 6% of this on train fares.

        How much does he spend on train fares?                                                             (3)

   (b)  David drinks 460 ml of juice from a bottle which contains 2 litres of
        orange juice.

        What percentage of the 2 litres does he drink?                                                     (3)

9  (a)  Write 48% as a fraction.                                                                            (2)

   (b)  Write $\frac{7}{25}$ as a decimal.                                                                  (2)

   (c)  Sheila has a £20 note. She spends $\frac{3}{8}$ of this on a book.

        How much does she spend on the book?                                                               (2)

   (d)  James eats 55% of a bunch of 40 grapes.

        How many grapes does James eat?                                                                     (2)

10 In 2010 the cost of Tommy's digital camera was £64 plus VAT (which
   was 17.5% of the basic price).

   To calculate the VAT, Tommy worked out

   (i)    a tenth of the £64 basic cost

   (ii)   half the answer to part (i)

   (iii)  half the answer to part (ii)

   (iv)   the sum of the answers to parts (i), (ii) and (iii).

   Find the total cost of the camera, including VAT at 17.5%.                                               (4)

11 (a) Rupert makes a coffee table for £30

He sells it, making a profit of 55%.

(i) How much profit does he make? (2)

(ii) For what price does Rupert sell the coffee table? (1)

(b) A coat, originally priced at £280, is sold for £168 in a sale.

(i) By how much has the price been reduced? (1)

(ii) Express this reduction as a percentage of the original price. (2)

12 (a) On the pet shop counter are some tins of dog food.

- $\frac{3}{5}$ are tins of *Wuff*.

- $\frac{1}{4}$ are tins of *Growl*.

- The rest are tins of *Yap*.

(i) What fraction of the tins are tins of *Yap*? (3)

There are 24 tins of *Wuff*.

(ii) How many tins of dog food are there altogether? (2)

(b) Rover eats $\frac{4}{5}$ of a tin of *Wuff* each day.

How many days will 12 tins of *Wuff* last? (2)

13 (a) Evaluate $\frac{3}{8}$ of 24 (2)

(b) Annie opens a box of sweets. She finds that:

- $\frac{1}{3}$ of them are mints

- $\frac{1}{4}$ of them are chocolates

- the rest are toffees.

What fraction of the sweets is toffees? (2)

(c) Tabitha eats $\frac{2}{3}$ of a tin of cat food each day.

How many full tins of cat food does Tabitha eat in November? (2)

14 (a) On this fraction strip, what fraction is shaded? (1)

(b) Copy this fraction strip and shade part of it to represent the result of the calculation $\frac{3}{5} + \frac{1}{5}$ (1)

(c) Evaluate:

(i) $\frac{1}{3} + \frac{3}{4}$ (2)  (ii) $\frac{3}{4} - \frac{1}{3}$ (2)

15 Evaluate:

(a) $\frac{1}{3} \times \frac{3}{4}$ (2)  (c) $1\frac{1}{3} + \frac{3}{4}$ (2)

(b) $\frac{1}{3} \div \frac{3}{4}$ (2)  (d) $2\frac{1}{3} - \frac{4}{5}$ (3)

16 Evaluate:

(a) $2\frac{3}{4} \times \frac{1}{3}$ (2)  (c) $2\frac{3}{4} + 3\frac{1}{3}$ (2)

(b) $2\frac{3}{4} \div \frac{1}{3}$ (2)  (d) $2\frac{2}{3} - 1\frac{3}{4}$ (3)

**Level 2** ●

17 Evaluate:

(a) $2\frac{3}{5} + 1\frac{2}{3}$ (3)

**Level 3** ■

(b) $\frac{2}{3}$ of $14\frac{1}{4}$ (3)

(c) $2\frac{3}{4} \div 1\frac{1}{3}$ (3)

**Level 3** ■

(d) $5\frac{3}{7} - 2\frac{8}{11}$ (3)

18 A DVD player is in a sale with 40% off its original price. The original price was £35

What is the sale price? (2)

19 In January 2008 the average house price in an area of Wales was £150 000

(i) If the price decreased by 8% over the following one-year period, what was the average house price in January 2009? (2)

(ii) If the house price then rose by 1% over the following 6 months, what was the average house price in the area in July 2009? Give your answer to the nearest £1000 (2)

(iii) Calculate the overall percentage fall in the average house price in the area between January 2008 and July 2009. Give your answer to 2 significant figures. (2)

20 (i) Change each of these fractions to a decimal, giving your answers to 4 decimal places.

   (a) $\frac{4}{7}$     (1)          (b) $\frac{8}{15}$     (1)          (c) $\frac{7}{12}$     (1)

   (ii) Write the fractions $\frac{4}{7}$, $\frac{8}{15}$ and $\frac{7}{12}$ in order of increasing size. (1)

   (iii) What is the difference between the smallest and the largest of these fractions? Give your answer as a fraction in its lowest terms. (3)

21 The diagram shows woodlice in a choice chamber.

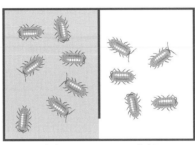

dark            light

   (i) How many woodlice are in the dark? (1)

   (ii) What is the fraction of the total number of woodlice that are in the dark? (1)

   (iii) What is the ratio 'woodlice in the dark' to 'woodlice in the light'? (2)

After an hour, three woodlice move from the light into the dark.

   (iv) What is the new ratio, in its simplest form, 'woodlice in the dark' to 'woodlice in the light'? (1)

22 A school has 120 pupils. $\frac{3}{5}$ of the pupils are boys.

   (a) What is the number of:

      (i) boys (2)

      (ii) girls? (1)

   (b) What is the ratio, in its simplest form, of:

      (i) girls to the total number of pupils (2)

      (ii) girls to boys? (2)

**23 (a)** Phil and Alex share 65 sweets in the ratio 8:5

How many sweets does Alex receive? (3)

**(b)** A cook uses 3 kilograms of sausages to make a meal for 27 people.

How many kilograms of sausages should she use to make a meal for 45 people? (2)

**24 (a)** Alice, Brendan and Corrie share a sum of money in proportion to their ages.

Alice is 12 years old and receives £6

If Brendan is 10 years old and Corrie is 13 years old, what is the total sum of money shared out between the three children? (3)

**(b)** The sizes of the angles of a quadrilateral *ABCD* are in the ratio 2:7:7:2

**(i)** Calculate the size in degrees of each angle. (4)

**(ii)** What shape is *ABCD*? (1)

**25** Sweets are shared between Morag and Hamish in the ratio 2:5

**(i)** If Morag receives 8 sweets, how many will Hamish receive? (2)

**(ii)** If Morag receives *m* sweets, how many will Hamish receive? (2)

**(iii)** There are 24 sweets in a bag of *Jollys*.

If as many sweets as possible from the bag are shared out in the ratio 2:5, as before:

**(a)** how many will Morag and Hamish each receive, and how many will be left? (2)

**(b)** how many *Jollys* should Hamish eat if the ratio of *Jollys* remaining between Morag and Hamish is to change to 3:2? (2)

# 2 Calculations

## 2.1 Mental strategies

In this section the questions *must* be answered:

- *without* using a calculator ☒
- *without* using any measuring instruments
- *without* making any written calculations.

These questions cover all areas of the syllabus; they are *not* grouped by topic or graded in difficulty.

The questions are printed as they would be read, using words rather than numerals in most cases. *Hearing* the questions read by someone else involves remembering the important details. In these questions you have the opportunity to re-read the questions as necessary, which makes life easier in many cases! The questions should be answered as quickly as possible and an average of about 10 seconds should normally be sufficient for each part, or about a minute for each numbered question of 5 parts.

For many questions, a variety of strategies can be used.

1  (a) I invest six hundred pounds at six per cent interest. How much interest do I earn in a year? (1)

   (b) Sarah had played in four hockey matches before today, and the figures show the number of goals she scored in each match. How many goals does she need to score today to make her mean number of goals three per match? (1)

   | 2 | 3 | 4 | 1 |
   |---|---|---|---|

   (c) A recipe says that ten kilograms of flour will make twenty loaves of a certain size. How many kilograms of flour would be needed to make thirty loaves of the same size? (1)

   (d) On a trip to London, Louise spends twelve pounds twenty pence on the train, ten pounds on a taxi, one pound thirty pence on a bus and two pounds seventy on the underground. How much does the journey cost her altogether? (1)

   (e) The information shows the charges for parking in a car park for different lengths of time.

   | 0–2 hours | 80p |
   |---|---|
   | Each extra hour or part hour | 80p |

   If I arrive at eleven thirty in the morning and leave at four fifteen the same day, how much do I pay to park? (1)

2   (a)  The calculation here is used to find the total cost in pounds of some items in a sale.

$$\frac{12 \times 16}{0.8}$$

Calculate the total cost. (1)

(b)  Tom is twelve years old. Four years ago his sister was twice his age. How old is she now? (1)

(c)  An A5 sheet of paper, half the size of this page, has an area of about three hundred square centimetres. What approximate area *in square metres* could be covered by one hundred sheets of A5 paper? (1)

(d)  How many cubical boxes of edge half a metre can be fitted into a case measuring three metres by two metres by one metre? (1)

(e)  The circumference of Terry's cycle wheel is two hundred and fifty centimetres. Estimate, to 1 significant figure, how many times the wheel goes round while she cycles ten kilometres. (1)

3   (a)  In Rugby Union, a converted try is worth seven points. A penalty is worth three points. In a match, a team scores three converted tries and two penalties. What is the team's total score? (1)

(b)  The total cost of a coach trip for thirty people is four thousand two hundred pounds. How much will each person pay if the cost is divided equally between them? (1)

(c)  Tiles are sold in packs of ten and cost eight pounds per pack. Bill estimates that a bathroom requires two hundred tiles. How much will the tiles cost? (1)

(d)  A pudding which feeds four people requires three hundred and twenty millilitres of milk. How much milk (in ml) is required if the pudding is to feed six people? (1)

(e)  Find the value of seven squared minus eight. (1)

4   (a)  A fire destroyed thirty out of the one hundred and twenty paintings in a gallery. What percentage of the original number of paintings was destroyed by the fire? (1)

(b)  The masses of three packets are ninety-five grams, forty-five grams and seventy grams. What is the mean mass of the packets? (1)

(c)  The cost to enter an adventure centre is twelve pounds per person. There is a reduction of one quarter for parties of ten or more. What is the total cost for a group of ten people to go to the adventure centre? (1)

(d) Calculate, in hectares, the area of the rectangular field shown.  (1)

400 m

150 m

Diagram not to scale; 1 hectare = 10 000 square metres.

(e) Mary needs to be in Puddleville by nine forty-five a.m. What is the latest time she can catch the bus in Grimvale?  (1)

| Grimvale depart | 0720 | 0810 | 0900 |
| Middleton arrive | 0805 | 0855 | 0945 |
| Puddleville arrive | 0835 | 0925 | 1015 |

5 (a) From the list of scores, write down the modal score.  (1)

| 6 | 5 | 8 | 3 | 3 | 9 | 3 | 6 | 9 |

(b) At midday the temperature was eleven degrees Celsius. By midnight it had dropped to negative three degrees Celsius. By how many degrees had the temperature fallen?  (1)

(c) Janet bought forty apples. She gave away forty percent of them. How many apples did Janet give away?  (1)

(d) Write down the total cost of six shirts which cost nine pounds ninety-nine pence each.  (1)

(e) Look at the expression:

$5p - 4$

What is the value of the expression when $p$ is six?  (1)

6 (a) There are twenty-six children in a youth group. Fourteen of them are boys. What is the probability that a person selected at random from the group is a girl?  (1)

(b) Look at the diagram.

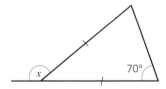

70°

$x$

What is the value in degrees of angle $x$?  (1)

(c) In a box of chocolates, the ratio of hard centres to soft centres is two to three. If there are eight hard centres, how many chocolates are there in the box?  (1)

(d) Choose three of these numbers that total one hundred. (1)

| 58 | 32 | 80 | 25 | 44 | 38 | 31 | 40 |

(e) In the long jump, Ian's first jump was four metres and sixty-seven centimetres. His second jump was five metres and six centimetres. By how many centimetres had he improved? (1)

7 (a) Work out six multiplied by five multiplied by four. (1)

(b) During a flu epidemic, one-fifth of a class of twenty children was absent. How many children in the class attended school that day? (1)

(c) Janet buys two magazines at ninety-five pence each. How much change does she receive from five pounds? (1)

(d) Terry's mother celebrated her fortieth birthday in 2007. In which year was she born? (1)

(e) Work out the value of $x$ squared plus four when $x$ equals five. (1)

8 (a) A bag contains twenty-five kilograms of compost. How many bags are delivered when Fred orders half a tonne of compost? (1)

(b) Jack came fourth in a running race with a time of three minutes and six seconds. The winner was thirteen seconds faster. What was the winning time? (1)

(c) A three hundred and five gram box of chocolates contains fifteen chocolates. Estimate the mean mass of a chocolate in the box. (1)

(d) A candle will burn for approximately fifteen hours. Robbie burns one of these candles for two hours each evening. If he lit it for the first time on December the first, on which date would he expect the candle to burn out? (1)

(e) The diameter of a two pence coin is two point five centimetres. Two pounds worth of two pence coins are laid in a straight line, just touching each other.

What is the length, in metres, of the line of coins? (1)

9 (a) A film begins at eighteen hundred hours and lasts one hundred and fifteen minutes. At what time does it finish? (1)

(b) Fifty-seven pounds is shared equally between three sisters. How much money does each sister receive? (1)

(c) Write down the value of three $x$ minus four when $x$ equals seven. (1)

(d) Work out zero point three four multiplied by two hundred. (1)

(e) The flag in the box on the left is rotated through one hundred and eighty degrees. Which number flag shows its new position? (1)

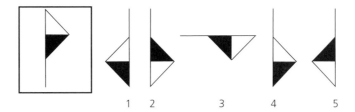

1    2        3        4        5

10 (a) Sixteen children in a class of twenty-eight play the piano. What is the probability that a child, chosen at random from the class, does not play the piano? (1)

(b) A kilt normally costs one hundred and ten pounds. The price is reduced by ten per cent in a sale. Work out the sale price. (1)

(c) On the map shown, one centimetre represents twenty metres. Estimate the distance in metres along the road from P to Q. (1)

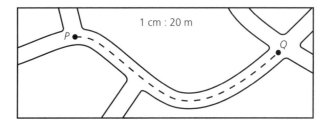

(d) A clock gains one minute every two hours. If the clock shows the correct time at midday on Monday, what time will it show at midday on Tuesday? (1)

(e) Write down three different whole numbers, each less than nine, whose product is thirty-six. (1)

11 (a) Alex buys a chocolate bar for forty-eight pence and a tube of mints for thirty-two pence. How much does he spend altogether? (1)

(b) Lucy celebrated her thirteenth birthday in 2008. In which year was she born? (1)

(c) What is a quarter of thirty-six? (1)

(d) A magazine costs two pounds twenty-five pence. What is the cost of four magazines? (1)

(e) Julie uses three hundred and seventy grams from a one kilogram bag of flour. How many grams of flour are left in the bag? (1)

**12 (a)** What is ten percent of three hundred and forty-five pounds? (1)

**(b)** Through how many degrees does the minute hand of a clock turn in fifty minutes? (1)

**(c)** Complete this fraction so that it is equal to three-fifths. (1)

$$\frac{?}{30}$$

**(d)** How many minutes are there between twenty forty-five hours and twenty-one thirty-five hours? (1)

**(e)** Jenny jogs at a steady speed of six kilometres per hour. How far does she jog in three and a half hours? (1)

**13 (a)** The spinner has six equal sections. What is the probability that it lands on a one? (1)

**(b)** Five oranges cost one pound eighty pence. What would be the cost of three of the same oranges? (1)

**(c)** These triangles are congruent. What is the size of angle $x$? (1)

**(d)** How many minutes are there in five days? (1)

**(e)** In a bowl of twenty apples and oranges, Terry counted eight apples. Write the ratio of apples to oranges in its simplest form. (1)

**14 (a)** Louise lives three kilometres from the station. She walks at four kilometres per hour. What is the latest time she can leave home to catch a train that leaves at nine o'clock in the morning? (1)

**(b)** A photograph measuring six centimetres by nine centimetres is enlarged by scale factor two. What is the perimeter of the enlarged photograph? (1)

**(c)** Find the value of zero point zero two multiplied by zero point five. (1)

**(d)** A tennis coach earns twenty-two pounds per hour. Last week he earned six hundred and eighty-two pounds. For how many hours did he work? (1)

(e) The square and the triangle have the same area. Find the distance *h*. (1)

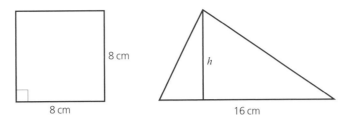

15 (a) Katie is thinking of two prime numbers. The sum of the numbers is eighty and the difference between them is six. What are the numbers? (1)

(b) Ben is thinking of three numbers. The mean of the numbers is thirty and the largest number is forty-six. The difference between the two smaller numbers is two. What is the smallest number? (1)

(c) Sarah is thinking of a two-digit number, *A*. She reverses the digits to get a new number, *B*, which is thirty-six more than *A*. What is Sarah's number? (1)

(d) Jack has forgotten his four-digit PIN but he *can* remember that the digits are all different primes and the first digit is two. How many attempts would he need before being sure to enter the correct PIN? (1)

(e) Olivia is thinking of a positive number. She squares it and then subtracts her original number to get thirty. What is her number? (1)

16 (a) Clare is thinking of a number between ten and twenty which is a multiple of four and a factor of forty-eight. List the possibilities. (1)

(b) The ages of Tom, Dick and Harry are consecutive prime numbers and the range of their ages is ten years. It will be twenty-four years before their ages are again all prime. How old is the youngest now? (1)

(c) Jo is twice as old as she was six years ago. How old will she be in three years' time? (1)

(d) Liam has chosen two fractions. The product of the fractions is the same as their difference. What are the fractions? (1)

(e) Bella thinks of a negative number. She squares it and adds twice her original number. She gets eight. What is her number? (1)

## 2.2 Written methods

*In this section the questions must be answered without using a calculator.* ✖

It is expected that all working is clearly set out. This will help you to avoid errors and gain more marks.

For many questions, a variety of strategies can be used.

Remember that in many cases, the correct 'answer' is less important than the 'working'.

An estimate before doing the calculation and a check afterwards will help to eliminate careless errors leading to 'ridiculous' answers.

1  Evaluate:

    (a)  5.75 + 0.649    (2)     (c)  5.75 × 0.5    (2)

    (b)  5.75 − 0.649    (2)     (d)  5.75 ÷ 0.5    (2)

2  Evaluate:

    (a)  6.47 + 0.34 + 2.08    (2)     (c)  13.9 × 0.4    (2)

    (b)  8.45 − 6.28    (2)     (d)  8.35 ÷ 5    (2)

3  Evaluate:

    (a)  31.4 + 6.7    (2)     (c)  3.88 × 4    (2)

    (b)  31.4 − 6.7    (2)     (d)  3.88 ÷ 4    (2)

4  Evaluate:

    (a)  37.9 + 8.4    (1)     (c)  33 − (⁻7)    (1)

    (b)  3000 − 497    (2)     (d)  3.5 × 0.24    (2)

5  Evaluate:

    (a)  13.82 + 6.48    (2)     (c)  6.48 × 3    (2)

    (b)  13.82 − 6.48    (2)     (d)  6.48 ÷ 3    (2)

6  (i)  Simplify the following fractions:

      (a)  $\dfrac{5+7}{14}$    (2)     (b)  $\dfrac{5-3}{6}$    (1)

   (ii)  Calculate the product of the two fractions in part (i).    (2)

7  (a)  Multiply 295 by 24    (2)

    (b)  Find the remainder when 1395 is divided by 8    (2)

8   Find the value of

(a)  3 + ⁻4 − ⁻2                    (2)            (b)  (⁻3 + ⁻4) × ⁻2              (2)

9   (a)  Sebastian bought 8 bargain DVDs which each cost £3.75

     How much did he pay altogether?                                          (2)

    (b)  Jemima bought 4 copies of the school choir DVD, to give as Christmas
         presents, at a total cost of £47.80

         What was the price of each DVD?                                      (2)

    (c)  Charlotte bought one rare DVD which cost £25.45

         How much change did she receive from a £50 note?                     (2)

    (d)  Tony spent £18.60 on two DVDs. One DVD cost twice as much as the
         other. What were the prices of the two DVDs?                         (3)

10  Evaluate the following:

    (a)  6.74 + 4.89          (2)            (c)  6.74 × 7                (2)

    (b)  6.74 − 4.89          (2)            (d)  4.89 ÷ 3               (2)

11  Evaluate:

    (a)  0.47 × 50   (2)      (b)  700 ÷ 0.2   (2)      (c)  70 − 14 ÷ 7   (2)

12  (i)  Calculate:

    (a)  4.6 + 5.6 ÷ 5                                                        (2)

    (b)  1.8 + 2.8 × (3.8 − 4.8)                                             (2)

    (ii)  What is the difference between your two answers in part (i)?        (2)

## 2.3 Calculator methods

*In this section a calculator is essential for most questions.*

In a few questions a calculator should be used for only part of the question and
the non-calculator part is clearly indicated by this symbol. **✗**

In most cases it is important to write more than just the 'answer'. Say what you
are doing.
Remember that the calculator will respond faultlessly to the instructions given to
it so it is very important that you give it the appropriate instructions!
It is a good idea to know roughly what the calculator answer is likely to be
before you start and a check after the calculation is always a good idea.
Remember that different calculators work in different ways and you should have
absolute confidence in your own calculator and your ability to use it.
Unless instructed otherwise, write all the figures shown in the calculator display
first and then write the answer to 3 significant figures.

1  (i)  Writing down all the figures shown on your calculator, find the value of

$$\frac{13.5 - 8.7}{12.6}$$  (2)

   (ii)  Write your answer to part (i) correct to 3 significant figures.  (1)

   (iii)  Write your answer to part (i) correct to 1 decimal place.  (1)

2  (a)  $\frac{354.7}{29.9 + 7.8}$

   *Without using a calculator*, and showing all your working:

   (i)  rewrite the calculation shown, giving each number correct to
        1 significant figure  (2)

   (ii)  evaluate your answer to part (i) giving your answer to 1 significant
         figure.  (1)

   (b)  (i)  Using a calculator and writing down all the figures displayed, evaluate

        $$\frac{492.7}{37.2 + 8.9}$$  (1)

   (ii)  Write your answer to part (b) (i) correct to 2 significant figures.  (1)

   (iii)  Write your answer to part (b) (i) correct to 1 decimal place.  (1)

3  (a)  (i)  Writing down all the figures shown on your calculator, evaluate

        $$\frac{41.3 + 28.7}{7.15 \times 11.3}$$  (2)

   (ii)  Write your answer to part (i) correct to 2 significant figures.  (1)

   (iii)  Write your answer to part (i) correct to 3 decimal places.  (1)

   (b)  *Without using a calculator*, and showing all your working:

   (i)  rewrite the calculation $\frac{8109}{42 \times 19}$ giving each number correct to
        1 significant figure  (1)

   (ii)  evaluate your answer to part (b) (i).  (2)

4  (a)  (i)  Writing down all the figures on your calculator display,
            find the value of

        $3.61 \times 3.52 + 6.715$  (2)

   (ii)  Write your answer to part (a) (i) correct to 1 decimal place.  (1)

   (b)  From a strip of wood, Edward saws eight identical pieces, each 5.6 cm long.

        What is the total length of wood that Edward saws off?  (2)

5 (a) (i) Using a calculator and writing down all the figures displayed, work out

$$\frac{57.2 \times 3.4}{9.2 - 4.7}$$ (2)

(ii) Write your answer to part (i) correct to 3 decimal places. (1)

(iii) Write your answer to part (i) correct to 2 significant figures. (1)

 (b) *Without using a calculator*, and showing all your working, estimate the value of this expression, giving your answer correct to 1 significant figure.

$$\frac{90.4 - 39.9}{51.6 - 1.95}$$ (3)

6 (i) Writing down all the figures shown on your calculator, evaluate

$$\frac{357}{13.4 \times 10.8}$$ (2)

(ii) Write your answer to part (i) correct to 2 decimal places. (1)

7 This calculator display was the result of finding the square root of a whole number, $x$.

| 6.708203932 |

(i) Write the number in the display to 3 significant figures. (1)

(ii) What was the original number, $x$? (1)

8 (i) Evaluate, writing down all the figures shown on your calculator,

$$\frac{20\,945 \times 32\,097}{694 \times 989}$$ (2)

(ii) Write your answer to part (i) correct to 2 significant figures. (1)

9 (i) Writing down all the figures shown on your calculator, evaluate

$$\frac{30.1 \times 1.95}{61.5 - 43.1}$$ (2)

(ii) Write your answer to part (i) correct to 3 decimal places. (1)

(iii) Write your answer to part (i) correct to 2 significant figures. (1)

# Problem solving

## 3.1 Reasoning about numbers or shapes

The questions in this section cover many areas of the syllabus.

You will

- make use of your knowledge and experience
- make observations
- find it helpful to ask questions 'What if …?'

*In this section you should not need to use a calculator.*

For many questions, a variety of strategies can be used.

1   A plant produces two flowers in its first year. Every year after that, it produces one more flower than in the previous year, as shown in this diagram.

year one                    year two                    year three

(i)   Copy and complete this table to show the number of flowers in each year.                                                                              (2)

| Year | Number of flowers |
|------|-------------------|
| 1    | 2                 |
| 2    | 3                 |
| 3    | 4                 |
| 4    |                   |
| 5    |                   |

(ii)  How many flowers will the plant have in its seventh year?                    (2)

(iii) Assuming that the plant continues to grow at the same rate, during which year will the plant produce 12 flowers?                                        (2)

2   The diagram shows a jigsaw puzzle cut from a rectangular piece of wood measuring 12 cm by 9 cm. All the 'knobs' on the pieces are of equal size. Disregarding the knobs, each piece is square and of equal size.

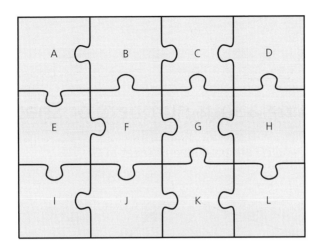

(i)   Taking the knobs into account:

    (a) which piece has the largest area (1)

    (b) which piece has the smallest area (1)

    (c) which pieces have the same area as piece E? (1)

(ii)   If the area of piece C is 8.4 cm², find the area of piece F. (3)

3   Here are the first three patterns in a sequence.

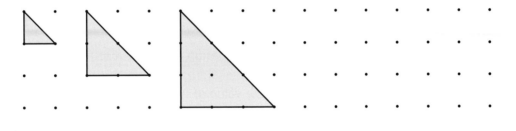

(i)   On a centimetre square dotted grid, draw the fourth pattern. (1)

(ii)   Copy and complete this table for patterns 1 to 4. (3)

| Pattern number | 1 | 2 | 3 | 4 |
|---|---|---|---|---|
| Number of dots on perimeter | 3 | | | |
| Number of dots inside pattern | 0 | 0 | | |
| Total number of dots | 3 | | | 15 |

(iii)   How many dots are there on the perimeter of pattern 6? (1)

(iv)   How many dots are there inside pattern 7? (2)

(v)   What is the total number of dots in pattern 8? (2)

4  (a)  What is the sum of the interior angles of a regular hexagon?          (2)

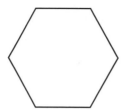

(b)  *ABCDEF* is a hexagon.

BC and ED are produced to meet at G.

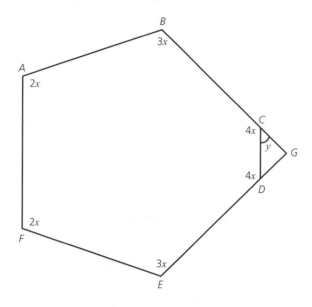

(i)   Write down, in terms of $x$, the sum of the interior angles of the
      hexagon *ABCDEF*.                                                        (1)

(ii)  Form an equation and solve it to find the value of $x$.                  (2)

(iii) Find the value of $y$.                                                   (2)

(iv)  Find the size of angle *BGE*.                                            (2)

Level 2

5  *A*, *B* and *C* are three vertices of a regular polygon.

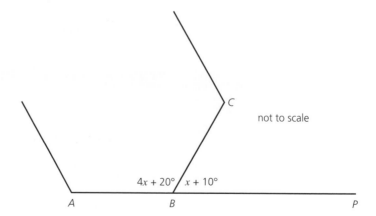

The line *AB* is extended to *P*.

(i) Write down an equation in $x$ linking the angles at *B*. (1)

(ii) Solve the equation to find $x$. (3)

(iii) State the value of the exterior angle *PBC*. (1)

(iv) Find the number of sides of the regular polygon. (2)

6 Many numbers can be expressed as the sum of two different prime numbers, for example $15 = 2 + 13$

(i) Express the following numbers as the sum of two different prime numbers:

(a) 16 (1) (b) 28 (2) (c) 42 (2)

(ii) Find the six different pairs of prime numbers which give a total of 60 (1), (1), (1), (1), (1), (1)

7 (a) A square has area 64 cm².

What is the perimeter of the square? (2)

(b) A square has perimeter 28 cm.

What is the area of the square? (1)

(c) The length of this rectangle is four times its width.

If the width of the rectangle is $x$ cm, write down an expression, in terms of $x$, for:

(i) the length of the rectangle (1)

(ii) the perimeter of the rectangle. (1)

The perimeter of the rectangle is 70 cm.

(iii) Form an equation and solve it to find the value of $x$. (1)

(iv) What is the area of this rectangle? (1)

(d) In another rectangle, the length is 7 cm more than the width.

The perimeter is 54 cm.

Use any suitable method to find the area of this rectangle. (4)

8  Look at the number puzzle set out here.

We wish to find the value of A and the value of B.

|     |     |     |     | Row total |
|-----|-----|-----|-----|-----------|
| A   | A   | A   | B   | 27        |
| A   | A   | B   | B   | ___       |
| A   | B   | B   | B   | ___       |

Column total ___  ___  24  ___  ___

(i)  Write equations for:

    (a)  the top row of the puzzle                                               (2)

    (b)  the third column of the puzzle.                                         (2)

(ii)  Solve the two equations in part (i) to find the values of A and B.          (3)

(iii)  Copy and complete the puzzle by writing the remaining totals.             (3)

(iv)  Several positive integers can be written as a sum of whole numbers
    of As and Bs, for example $A + B$; $2A + B$

    Which integers between 30 and 40 inclusive can be written in this way?     (3)

9  A slug is moving along a path. Each minute it moves 3 cm further than it moved
during the previous minute. It moved 2 cm during the first minute.

(i)  Write down the next four terms of the sequence starting with 2,
    to show the number of centimetres the slug moves during each of
    the first five minutes.                                                   (2)

(ii)  How far does the slug move during the eighth minute?                        (1)

(iii)  What is the total distance that the slug will have moved
    after eight minutes?                                                      (2)

(iv)  Write down the 20th term of the sequence in part (i).                       (2)

10  In this investigation we see what happens when we apply a rule over and over
again to create a sequence.

We will start with 85 and apply the rule:

Square the units digit and add the tens digit.

So the sequence runs:

| First term           |            | 85 |
|----------------------|------------|----|
| then $5 \times 5 + 8 =$ | $25 + 8 =$ | 33 |
| then $3 \times 3 + 3 =$ | $9 + 3 =$  | 12 |
| then $2 \times 2 + 1 =$ | $4 + 1 =$  | 5  |
| then $5 \times 5 + 0 =$ | $25 + 0 =$ | 25 |

(i) Find the next 18 terms in the sequence, showing your working, as in the previous table. A few terms are provided as a check. (5)

| | | | |
|---|---|---|---|
| Sixth term | | | |
| Seventh term | | | 51 |
| Eighth term | | | |
| | | | |
| | | | 39 |
| | | | |
| Twelfth term | | | |
| | | | |
| | | | |
| | | | 31 |
| | | | |
| | | | |
| | | | |
| | | | |
| | | | |
| Twentieth term | | | |
| | | | |
| | | | |
| | | | |

(ii) What do you notice? (2)

(iii) Write down the next five terms if you start with

(a) 37 (2)     (b) 54 (2)

**11** (i) Here is a rectangle. It is three times as long as it is wide.

$x$

(a) If the width is $x$ cm, write down an expression, in terms of $x$, for the perimeter of the rectangle. (2)

(b) If the perimeter is 40 cm, what is the area of the rectangle? (2)

(ii) In another rectangle, the length is 2 cm more than the width. The perimeter of this rectangle is also 40 cm.

Find the area of this rectangle. (3)

**12** Look at these fractions:

$$\frac{2+3}{9} \qquad \frac{1+2}{5} \qquad \frac{3+4}{11}$$

(i) Which fraction has the largest value? (2)

(ii) Calculate the difference between the largest and the smallest fractions. (2)

13 Three major male roles in the school Shakespeare play are to be decided. The parts are Macbeth, Macduff and Banquo. Four boys, Peter (P), Quentin (Q), Rachid (R) and Sam (S) are keen to be chosen.

Since there are only three major parts, one boy will be left out. Peter says he would play only Macbeth, but the other three boys would be happy with any part.

(i) Copy this table and list the different ways in which the three major parts could be allocated. Two are done for you. (3)

| Macbeth | Macduff | Banquo |
|---|---|---|
| P | Q | R |
| | | |
| | | |
| | | |
| | | |
| | | |
| Q | R | S |
| | | |
| | | |
| | | |
| | | |
| | | |

Each of the possible combinations is written on a piece of paper. The pieces are put in a hat and one is chosen at random.

(ii) What is the probability that Quentin is chosen to play a part? (2)

(iii) If Quentin is chosen to play the part of Banquo, what is the probability that Peter is left out altogether? (2)

14 Gina builds a pyramid using one pound coins.

She starts with a layer of coins arranged as shown (right) with the coins touching.

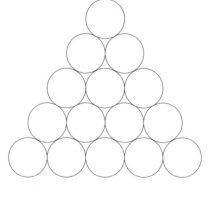

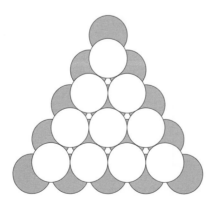

She continues with a second layer, as shown (left), with the centres of the coins above the centres of the gaps in the layer below.

(i) In order to complete this pyramid:

    (a) how many more layers will Gina add (1)

    (b) how many more coins will Gina add (1)

    (c) how many coins will Gina have used altogether? (1)

Georgie thinks that it would be a good idea to build a similar pyramid with ten layers but she is unsure how many coins there should be in the bottom layer.

Gina suggests working backwards, starting with the top layer, and prepares a table.

| Layer | Number of coins on an 'edge' | Number of coins in a layer | Total number of coins used so far |
|---|---|---|---|
| 1 (top) | 1 | 1 | 1 |
| 2 | 2 | 3 | 4 |
| 3 | 3 | 6 | |
| 4 | | | |
| 5 | | | |
| 6 | | | |
| 7 | | | |
| 8 | | | |
| 9 | | | |
| 10 | | | |

(ii) Copy and complete the table. (5)

Georgie gets the coins from the bank and builds the pyramid with ten layers.

A pound coin has

● mass           9.5 grams

● diameter      22.5 mm

● thickness     3.15 mm

(iii) Calculate:

    (a) the height of the pyramid (in cm) (2)

    (b) the mass of the pyramid (in kg). (2)

29

**15** The diagram shows a grid pattern of paths in a maze.

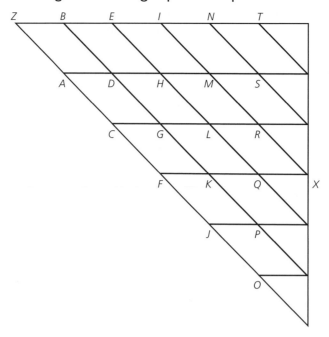

Visitors must enter the maze at *Z* and they must keep moving east or south-east only.

The exit is *X*.

Following the rules, there is only one way of reaching point *A* but there are two ways of reaching point *D*.

**(i)** How many ways are there of reaching points

    **(a)** *G*     (1)         **(b)** *H*     (1)         **(c)** *I*?     (1)

**(ii)** Making use of your answers to part (i), find the number ways of reaching points:

    **(a)** *J*     (1)         **(b)** *K*     (1)         **(c)** *L*.     (2)

**(iii)** How many possible routes are there from *Z* to *X*?     (4)

*Show all your reasoning and working clearly.*

**16** Morag is investigating the differences between consecutive cube numbers.

She calculates the **digital root** for each difference.

For example, the digital root of 169 would be found as follows:

$1 + 6 + 9 = 16; 1 + 6 = 7$

The digital root (which is a single digit) of 169 is 7

(i) Copy and complete the table. (8)

| $N$ | $n^3$ | $(n-1)^3$ | $n^3 - (n-1)^3$ | digital root of $(n^3 - (n-1)^3)$ |
|---|---|---|---|---|
| 1 | 1 | 0 | 1 | 1 |
| 2 | 8 | 1 | 7 | 7 |
| 3 | 27 | 8 | 19 | 1 |
| 4 | 64 | | | |
| 5 | | | | |
| 6 | | | | |
| 7 | | | | |
| 8 | | | | |
| 9 | | | | |
| 10 | | | | |

(ii) What do you notice about

(a) the differences between the values in the $n^3 - (n-1)^3$ column (1)

(b) the sequence of numbers in the digital root column? (1)

(iii) Calculate the value of $3n^2 - 3n + 1$ when $n$ is:

(a) 5 (2)    (b) 8 (2)

(iv) Comment on your results in part (iii). (2)

(v) *Without using your calculator*, and showing all your working, find the value of $n^3 - (n-1)^3$ when $n$ is 100 (2)

17 (i) (a) Construct triangle *ABC* where *AB* is 10 cm, *AC* is 8 cm and *BC* is 6 cm. (2)

(b) Construct the bisectors of the three angles of triangle *ABC* and mark the point of intersection of the bisectors, *D*. (2)

(c) With centre *D*, open your compasses so that the pencil just touches *AB* at the nearest point and draw a circle – the **inscribed circle**. (1)

(ii) (a) Construct triangle *EFG* where *EF* is 10 cm, *EG* is 8 cm and *FG* is 6 cm. (2)

(b) Construct the perpendicular bisectors of the three sides of triangle *EFG* and mark the point of intersection of bisectors, *H*. (2)

(c) With centre *H*, open your compasses so that the pencil is just on *E* and draw a circle – the **circumscribed circle**. (1)

(d) What do you notice about *EF*? (1)

**18 (i) (a)** Construct two congruent obtuse-angled scalene triangles with sides of length 12 cm, 9 cm and 6 cm. (2)

**(b)** Construct the inscribed circle of one triangle. (2)

**(c)** Construct the circumscribed circle of the second triangle. (2)

**(ii) (a)** Construct two congruent equilateral triangles with sides of length 8 cm. (2)

**(b)** Construct the inscribed circle of one triangle. (2)

**(c)** Construct the circumscribed circle of the second triangle. (2)

**(iii)** Comment on your discoveries. (2)

**19** Sophia has *one* each of the following Roman numeral cards:

| M | D | C | L | X | V | I |
|---|---|---|---|---|---|---|

In each part of this question, write the number in

**(a)** Roman numerals

**(b)** ordinary numerals.

Sophia makes numbers by putting the cards side by side. For example, the smallest multiple of 12 she could make is LX (60).

| L | X |
|---|---|

Using Sophie's cards, write down

**(i)** the number closest to 200 (2)

**(ii)** the smallest square number after 1 (2)

**(iii)** the smallest cube number after 1 (2)

**(iv)** the smallest multiple of 13 (2)

**(v)** the largest factor of 144 (2)

# 3.2 Real-life mathematics

In this section many of the questions are concerned to some extent with money or with fractions, decimals and percentages.

The real-life situations may include:

- shopping – money, discounts
- cooking and baking – recipes
- eating and diets – calories
- medicines and health – measures
- growth – measurement, line graphs
- exercise and sports – times, speeds, scores
- holidays and travel – driving, exchange rates
- planning events – hiring equipment, fundraising
- gardening – weed killing, filling a pond
- DIY tasks – building a shed, decorating a room
- business and banking – VAT, profit and loss, interest rates
- design – tessellations, symmetry
- exploring – maps, scales
- model-making – scales, nets
- puzzles.

In this section the questions should be answered without using a calculator wherever possible. **X**

Questions where the use of a calculator is recommended or essential are indicated by

1 **(i)** *Estimate* the cost of 39.8 litres of petrol at 100.9 pence per litre.

Give your answer to the nearest pound. (2)

**(ii)** On a journey of 361 miles, Ethan's car uses 1 litre of petrol every 8 miles.

Estimate the number of litres of petrol that Ethan's car uses on the journey. (2)

2 **(a)** Jane buys a tube of toothpaste for £1.30 and a packet of tissues for 75 pence.

How much does she spend altogether? (2)

**(b)** David buys a tie for £14.49

He pays with a £20 note. How much change should he receive? (2)

**(c)** A kilogram of leeks costs £1.49

What is the cost of 9 kilograms of leeks? (2)

**(d)** Ben spends £3.60 on crusty rolls which cost 90 pence each.

How many rolls does he buy? (2)

3   In a sale, *Kilts4U* offers a discount of 40% off the marked price of all items.
    *Tartanforall* offers a reduction of $\frac{1}{4}$ off the marked price of all items.

    Hamish sees identical kilt outfits which are priced at £360 in *Kilts4U* and £300 in *Tartanforall*.

    (i)   How much would Hamish pay in *Kilts4U*? (3)

    (ii)  How much would Hamish pay in *Tartanforall*? (2)

    (iii) Which store offers Hamish the cheaper price and how much less
         would he pay? (2)

    (iv) Place the correct sign < or > in the statement

         40% of 360 _____ $\frac{1}{4}$ of 300 (1)

4   In a cycling race, Ben covered 18 km in 20 minutes.

    (i)   What was his average speed in:

        (a)  kilometres per hour (2)

        (b)  metres per second? (2)

    (ii)  Travelling at this average speed, how many kilometres would
         he travel in 12 minutes? (2)

5   (a)  Gina has her own recipe for making a special hot chocolate drink.
       For six people:

| chocolate | 300 g |
| milk | 900 ml |
| cream | 180 ml |

       Rewrite the recipe for ten people. (4)

    (b)  A box contains 36 biscuits.

       Una, Violet and Wendy share the biscuits in the ratio of 2:3:4

       How many biscuits does Wendy get? (3)

6   (a)  The ratio of the number of boys to the number of girls at Falcon
       Court is 5:4

       There are 52 girls.

       (i)  How many boys are there at Falcon Court? (2)

       (ii) Write, in lowest terms, the ratio of the number of boys to
          the total number of pupils. (2)

**(b)** A game uses the two 'rollers' A and B, shown here.

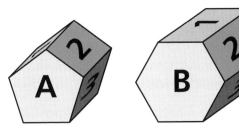

The rollers are both prisms with regular polygons as their end faces.

Name the regular polygons:

**(i)** A (1)      **(ii)** B (1)

**(iii)** Which roller will turn through the larger angle as it rolls from one numbered face to the next, and what will this angle be? (2)

**(iv)** Write, in its simplest form, the ratio of the size of the interior angle, $i$, to the size of the exterior angle, $e$, for the polygon of roller A. (2)

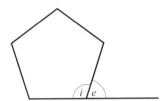

Roller A is numbered 1 to 5 and roller B is numbered 1 to 6

**(v)** What is the ratio of odd numbers to even numbers on roller A? (2)

**(vi)** What is the probability of scoring 5 with roller B? (1)

7   Uncle Tom has divided his stamp collection between his three nieces as follows:

- Angela has $\frac{1}{5}$ of the total

- Betty has $\frac{1}{4}$ of the total

- Clare has the rest.

**(i)** What fraction of the total does Clare have? (3)

Clare divides her share equally between her twin brothers, Dennis and Eli.

**(ii)** What fraction of the total does Dennis have? (2)

**(iii)** If Eli now has 660 of Uncle Tom's stamps, how many stamps were there in Uncle Tom's collection? (3)

8  (i)   Krishna bought 45 litres of petrol. He can remember that the total was a little more than £50 but he can't remember what the price per litre was.

Estimate the price per litre, in pence, of the petrol.  (3)

(ii)  The next time Krishna bought petrol, the price per litre was 108.1 pence and the total bill was about £42

Estimate the number of litres of petrol he bought.  (2)

9  (i)   Write the missing amounts in this shopping list.  (4)

> 3 kg of parsnips at 45 pence per kg cost ? pence
>
> 6 kg of potatoes at ? pence per kg cost 138 pence
>
> ? kg of sprouts at 68 pence per kg cost 204 pence

The total amount spent on vegetables is £?

(ii)  Find the total mass of vegetables bought.  (1)

(iii) Find the mean price per kilogram of the vegetables bought. Give your answer to the nearest penny.  (2)

10  Cheng buys a skateboard.

He makes a first payment of £40 and then pays £6.50 each month for the next 12 months.

How much does he pay altogether for the skateboard?  (3)

11  The angles of a triangle are in the ratio 1 : 3 : 5

Find the size of each angle.  (2)

12  Tom's grandfather can remember that 50 years ago his car averaged 40 miles to the gallon at 40 miles per hour and 30 miles to the gallon at 50 miles per hour.

(i)   He usually drove the 240 miles from Ogglington to Marshy at an average speed of 40 miles per hour.

How many gallons of petrol did he use for a trip from Ogglington to Marshy *and back* travelling at his usual speed?  (1)

(ii)  One day he was in a hurry and completed the return journey at an average speed of 50 mph.

How many more gallons would he have used than if he had travelled at the slower average speed?  (2)

Tom is amazed to learn that, 50 years ago, you could buy 4 gallons of petrol for just less than £1

(iii) What would be the approximate cost of petrol for a return trip from Ogglington to Marshy at the slower average speed?  (2)

13 (a) In a fun event, Ben ran 0.75 kilometres, cycled 3.5 kilometres and finally swam 200 metres.

What is the total distance, in metres, that Ben covered? (2)

(b) Pat has four bottles full of water.

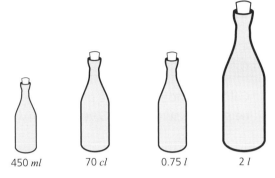

450 *ml*      70 *cl*      0.75 *l*      2 *l*

All the water is poured into an empty 6 litre container.

(i) What is the volume, in millilitres, of water in the container? (2)

(ii) How many litres of water should now be poured into the container to fill it? (1)

14 (a) George fills the tank of his car with 40.3 litres of petrol at 109.9 pence per litre.

Estimate the cost, in pounds, of the petrol, showing your working clearly. Give your answer to the nearest pound. (2)

(b) Gemma can run 200 metres in 36 seconds.

Calculate her speed in kilometres per hour. (2)

15 (i) Write 9 and 16 as products of prime factors. (2)

(ii) Use your answer to part (i) to write down the whole-number factors of 144 (9 × 16) which are multiples of 9. Two are given for you.

9, ..., ..., ..., 144 (2)

(iii) A 720 gram tray of mint tablet is made in a slab of 9 × 16 pieces.

(a) If the whole slab was shared equally between 48 children, how many pieces would each receive? (1)

(b) What fraction, in its lowest terms, of the whole slab would eight of these children have altogether? (2)

(iv) What is the total mass (in g) of four pieces of the tablet? (2)

(v) What percentage of the whole slab is 54 pieces of the tablet? (2)

**16 (a)** Express 36% as a fraction in its lowest terms. (2)

**(b)** Express 12.5% as a decimal. (2)

**(c)** At a warehouse in 2010, a sweatshirt was advertised at £18.40 excluding VAT (value added tax). VAT at 17.5% was added to the price at the checkout.

   **(i)** Calculate the price paid for the sweatshirt at the checkout. (2)

   Mr Jones paid the full price for a sweatshirt at the warehouse and sold it for £29.99 on his stall.

   **(ii)** Calculate the percentage profit made by Mr Jones when he sold the sweatshirt. Give your answer to 1 decimal place. (2)

**17** Pat is taking part in a 150 mile sponsored cycle ride.

**(i)** He cycles the first 65 miles in 5 hours.

   What is his average speed in miles per hour? (2)

**(ii)** Pat's friends, between them, sponsor him at a rate of 85 pence for each mile he cycles.

   **(a)** If he stops after 65 miles, how much sponsorship money does he earn? (1)

   In addition to his friends' sponsorship, Pat's mother sponsors him with a single payment of £20 if he completes the 150 miles.

   **(b)** How much more sponsorship money does Pat earn by finishing the ride, rather than stopping after 65 miles? (3)

**18** On a freeway in Western Australia Bert drives at a steady 105 km/h.

One day he drives from 9 o'clock in the morning until 7 o'clock in the evening, with a one-hour stop for lunch.

How many kilometres does he drive? (3)

**19** A circular table has a radius of 72 cm.

**(i)** Calculate the area of the table top, giving your answer to the nearest 10 square centimetres. (2)

Five circular place mats, each of diameter 22 cm, are placed on the table.

**(ii) (a)** Calculate the area of one place mat, giving your answer to 2 significant figures. (1)

   **(b)** Using your answers from parts (i) and (ii) (a), find the percentage of the area of the table top left uncovered when the five mats are placed on the table, giving your answer to the nearest 1 percent. (3)

Five people sit equally spaced round the table.

**(iii)** Calculate the distance around the edge of the table available for each place setting, giving your answer to the nearest centimetre. (3)

**20** Mary's rectangular garden measures 18 metres by 14 metres. It has a quadrant-shaped flower bed, of radius 2 metres, in each corner as shown in the diagram.

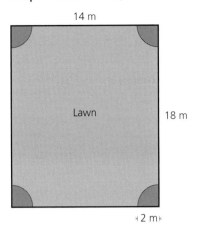

14 m

Lawn          18 m

⊣2 m⊢

    **(i)**    What is the area of one flower bed to the nearest square metre?    (3)

    **(ii)**    What area of the garden is lawn? Answer to 2 significant figures.    (2)

    **(iii)**    What percentage of the whole garden is lawn? Give your answer correct to 2 significant figures.    (3)

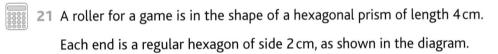 **21** A roller for a game is in the shape of a hexagonal prism of length 4 cm.

Each end is a regular hexagon of side 2 cm, as shown in the diagram.

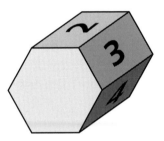

    **(i)**    Sketch the net of the roller.    (2)

    **(ii)**    **(a)**    A regular hexagon of side 2 cm is drawn on this centimetre squared grid. By counting squares, or otherwise, find the approximate area, in cm², of a hexagonal face of the roller. Explain clearly how you do this.    (3)

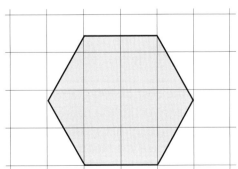

        **(b)**    Using your answer to part (ii) (a), calculate the surface area of the roller (cm²).    (2)

**22** Anna works at the checkout of a supermarket.

She earns £7.50 per hour for the first 30 hours in a week.

    **(i)**    How much is Anna paid for a 30-hour week?     (1)

Anna is paid £9.20 per hour for every extra hour she works over 30 hours.

    **(ii)**    One week Anna worked for a total of 41 hours.

         How much did she earn in that week?     (2)

    **(iii)**    Another week Anna earned £353.80

         How many hours did she work that week?     (3)

**23 (a)**    Angus buys a vase for £48

    He hopes to sell it, making a profit of 40%.

    **(i)**    For how much does he hope to sell the vase?     (2)

    Unfortunately nobody buys the vase, so Angus reduces his price by 5%

    **(ii)**    What is the new price of the vase?     (2)

    He sells the vase at the new price.

    **(iii)**    Express his profit as a percentage of his original cost price.     (2)

**Level 2** ●

**(b)**    Train fares were increased by 10% on 1 January.

    Maisie pays £198 for her ticket to London.

    How much would she have paid before 1 January?     (2)

**24** Regal Choice coffee is a blend of three types: A, B and C.

- A is bought in 20 kg bags each costing £180
- B is bought in 25 kg bags each costing £174
- C is bought in 22 kg bags each costing £210

    **(i)**    If the contents of one bag of each type were mixed together, what would be the cost of 100 grams of the mixture?     (2)

    **(ii)**    **(a)**    Find the cost of 100 grams of each of the three types of coffee. Give answers to the nearest penny.     (2)

         **(b)**    Regal Choice is a blend, by weight, of A, B and C in the ratio 1:3:2

         Calculate the cost of 500 grams of this mixture, giving your answer to the nearest penny.     (3)

25 Pizza Palace has a special offer:

> *Buy one full price,*
> *get another the same*
> *size half price.*
>
> _____
>
> Small pizza (20 cm)
> £3.30
>
> Medium pizza (25 cm)
> £5.20
>
> Large pizza (28 cm)
> £6.66

Angela wants to buy as much pizza as possible for her friends and has £10 to spend.

Assume that the pizzas are circular with diameters 20 cm, 25 cm and 28 cm, and they are of the same thickness.

(i) By considering the prices of the pizzas, list the possible combinations which Angela could afford to buy. *Ignore any possibilities which do not allow her to make use of the special offer.* (2)

(ii) By considering the areas of the circles, investigate how Angela should spend her money in order to get the maximum amount of pizza. (5)

26 On a journey across Europe, Brian and Brenda travel 286 kilometres at an average speed of 65 km/h.

(i) Find the time taken for this part of the journey. Give your answer in hours and minutes. (2)

After a rest of one hour and eight minutes, Brian and Brenda continue their journey, travelling at an average speed of 60 km/h, taking three hours and forty minutes.

(ii) Calculate the distance travelled on this part of the journey. (2)

(iii) Calculate the average speed for the complete journey, including the rest period. (3)

27 A DVD rental business automatically takes a fixed monthly card payment of £6.00 which covers an administration charge and the rental of up to two DVDs.

There is an additional charge of £2.50 for every extra DVD rented during the month. The £2.50 includes return first class postage since DVDs are booked online.

(i) During December, the Smith family rented a total of five DVDs.

What was the total charge for that month? (2)

**(ii)** In January the total payment charged to the card was £23.50

How many DVDs did the Smith family rent in January? (2)

**(iii)** In February, the Smith family rented only one DVD.

What was the charge for February? (1)

In March there was a 10% increase in all charges.

**(iv)** The Smith family's total charge for March was £23.10

How many DVDs did they rent in March? (3)

**28** Mr Williams has £1150 in his bank account and he has an agreed overdraft limit of £500

He pays £1399 for a new suite of furniture.

**(i)** Is he now in credit or debit at the bank, and by how much? (2)

He pays £400 into his bank account and checks his balance.

**(ii)** What is his balance now? (1)

**29** Harry has just bought, on his birthday, 1st January, a new mountain bike priced at £599.99

He used all of his birthday money to pay 10% of the cost price of the bike.

**(i)** How much money did he get on his birthday? (1)

The remainder of the cost is to be spread over 12 equal monthly payments.

**(ii)** How much will Harry pay each month? (3)

Harry earns £14 every week by delivering papers and he gets a weekly allowance (pocket money) of £12 from his parents.

**(iii)** How much, in a year (52 weeks), will Harry

**(a)** earn from the paper round (2)

**(b)** receive from his parents? (1)

**(iv)** How much, after paying for the bike, will Harry have available to spend on other things during the year? (3)

# 4 Algebra

## 4.1 Equations and formulae

In this section the questions cover the following topics:

- Terms and expressions
- Simplification; brackets
- Substitution
- Equations
- Formulae
- Modelling

Level 3 ■

- Inequalities
- Trial and improvement (this section is included for interest only)

Many questions cover several topics.

*In this section you should answer the questions without using a calculator except where indicated by* 🖩

1   Simplify the following expressions:

(i)   $4a + 3a$ (1)

(ii)   $4a \times 3a$ (2)

(iii)   $\dfrac{3a + a}{2}$ (2)

(iv)   $2a - 3a - a$ (2)

2   Simplify the following expressions:

(i)   $3a - 4b + b - 2a$ (2)

(ii)   $3a^2 \times 4a^3$ (2)

Level 2 ●

(iii)   $\dfrac{6a^2}{12a}$ (2)

3   (a)   Simplify

(i)   $3a - 4a + 2a$ (2)

(ii)   $\dfrac{8a + 12b}{4}$ (2)

(b)   Multiply out the bracket and simplify:

(i)   $2(x + y) - y$ (2)

(ii)   $3(a - b) - 2(a - 3b)$ (3)

4   (a)   Factorise completely: $5n - 15$ (2)

(b)   Multiply out the brackets and simplify: $2(p + 3q) - (p + 2q)$ (3)

5  (a)  Simplify, by collecting like terms: $24m + 3n - 7n - 19m$  (2)

   (b)  Factorise: $12a + 18b$  (2)

6  (a)  Multiply out the brackets and simplify the following expression:
        $2(5 - 3n) + 6(n - 3)$  (3)

   (b)  Factorise completely: $10u + 15$  (2)

7  (a)  Simplify, by collecting like terms: $2x + 3x^2 - 2 + 3x$  (2)

   (b)  Simplify: $2x^2 \times 3x^3$  (2)

Level 2 ●   (c)  Simplify: $\dfrac{18x^2}{6x^3}$  (2)

8  (a)  Simplify: $8y - 4 + 2y + 3$  (2)

   (b)  Remove the brackets and simplify: $2(x + 3y) + 3(x - 2y)$  (3)

9  (a)  Remove the brackets and simplify where possible:

        (i)   $7(2a + 5)$  (1)

        (ii)  $(5a - 3) - (a - 5)$  (2)

   (b)  Factorise:

        (i)   $6a + 10$  (2)

Level 2 ●        (ii)  $8ab - 12bc$  (2)

10 Simplify:

        (i)   $15p^2 - 17p^3 - 9p^2 + 4p^3$  (2)

        (ii)  $2(5 + 3h) - 4(2h - 5)$  (2)

Level 2 ●        (iii) $\dfrac{5w - 3w}{4w}$  (2)

11 (a)  Simplify:

        (i)   $2y^3 + 2y^3$  (1)

        (ii)  $2y^3 \times 2y^3$  (2)

Level 2 ●        (iii) $\dfrac{12y^2}{3y^6}$  (2)

   (b)  Multiply out the brackets and simplify: $3(2p - 5q) - 5(p + 3q)$  (3)

Level 3 ■   (c)  Factorise completely: $15a^2 + 25a$  (2)

12 (a)  Multiply out the brackets and simplify: $2(3a + 1) - 3(a - 2)$  (3)

   (b)  Factorise: $8a - 12b$  (2)

(c) Simplify:

(i) $3a^3 + 2a^2 - 2a^3 + 3a^2$ (2)

(ii) $4a^4 \times 3a^3 \times 2a^2$ (2)

**Level 2** ●

(iii) $3a^2 \div 2a^3$ (2)

**13 (a)** Factorise: $3 - 36y$ (2)

**Level 2** ●

(b) $\dfrac{18x^3}{(3x)^2}$ (2)

**14** Simplify the following algebraic expressions:

(i) $3ab + 2a^2 - 5ab + 7a^2$ (2)

(ii) $2(3m + 1) - 3(2m - 1)$ (3)

**Level 3** ■

(iii) $\dfrac{2x^2 + (3x)^2}{x}$ (3)

**15** When $a = 5$, $b = {}^-4$ and $c = 3$, find the values of the following:

(i) $a + c$ (1)      (iv) $b^2$ (2)

(ii) $a + b$ (1)      (v) $a(b - c)$ (2)

(iii) $a - b - c$ (2)

**16** When $x = 4$, $y = {}^-1$ and $z = 3$, find the values of:

(i) $3x - y$ (2)      (iii) $z(x + y)$ (2)

(ii) $3y^3$ (2)      (iv) $\dfrac{2xz}{y}$ (2)

**17** When $a = 4$, $b = {}^-3$ and $c = 1$, evaluate:

(i) $a - b - c$ (2)      (iii) $a^2 + b^2$ (2)

(ii) $ac^3$ (2)

**18** When $a = 3$, $b = 1$ and $c = {}^-2$, evaluate:

(i) $3a + 2b$ (2)      (iii) $(b - c)^2$ (2)

(ii) $b - c^2$ (2)

**19** If $p = 3$, $q = {}^-2$ and $r = 0$, evaluate:

(i) $p + q$ (2)      (iii) $p(r - q)$ (2)

(ii) $pqr$ (2)

**20** If $a = 4$, $b = {}^-3$ and $c = {}^-6$, find the values of:

(i) $bc$ (1)      (iii) $\dfrac{ab^2}{2c}$ (3)

(ii) $a(b - c)$ (2)

21 If $a = \frac{1}{2}$, $b = \frac{2}{3}$ and $c = \frac{^-3}{4}$, evaluate the following, giving your answers in their simplest form:

   (i)   $a + c$                                             (2)

   (ii)  $b(a + c)$                                      (2)

   (iii) $\dfrac{3b}{a}$                                        (2)

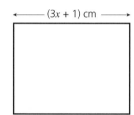

22 Consider the formula: $s = ut + \frac{1}{2}at^2$
When $u = 60$, $a = 8.8$ and $t = 6.5$, find:

   (i)   $ut$                 (1)                                (iii) $s$        (1)

   (ii)  $\frac{1}{2}at^2$             (1)

23 The length of a rectangle is $(3x + 1)$ cm.

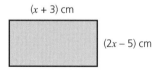

The width of the rectangle is $2x$ cm less than the length.

   (i)   Write down an expression, in terms of $x$, for the width of the rectangle. (1)

   (ii)  Use your answer to part (i) to find the perimeter of the rectangle in terms of $x$. Simplify your answer. (3)

The perimeter of the rectangle is 68 cm.

   (iii) Form an equation in $x$ and solve it. (2)

   (iv) Use your answer to part (iii) to find the length of the rectangle. (1)

   (v)  Calculate the area of the rectangle. (2)

24 A rectangular label has length $(x + 3)$ cm and width $(2x - 5)$ cm.

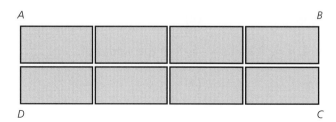

A sheet of these labels has eight labels arranged as shown by the rectangle *ABCD*. Ignore the gaps between the labels.

(i) Write down, in terms of $x$, simplifying your answer:

    (a) the length $AB$                                 (2)

    (b) the width $BC$.                                (2)

The perimeter of the sheet of eight labels, $ABCD$, is 52 centimetres.

(ii) Write down an equation in terms of $x$ and solve it.     (4)

25 In triangle $ABC$, $AB$ is $x$ cm.

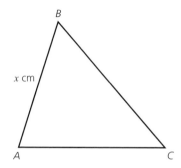

AC is 8 cm longer than $AB$.

(i) Write down an expression, in terms of $x$, for the length of $AC$.     (1)

$BC$ is four times the length of $AB$.

(ii) Write down an expression, in terms of $x$, for the length of $BC$.     (1)

(iii) Write down an expression, in terms of $x$, for the perimeter of triangle $ABC$.     (2)

The perimeter of the triangle is 32 cm.

(iv) Form an equation and solve it to find the value of $x$.     (3)

(v) What is the length of $AC$?     (1)

26 $ABC$ is a triangle and the length of side $AB$ is $x$ cm.

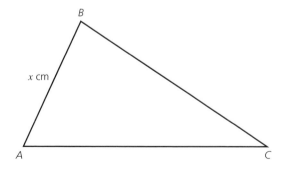

(i) The side $AC$ is three times as long as $AB$. Write down, in terms of $x$, the length of $AC$.     (1)

(ii) The side $BC$ is 5 cm longer than $AB$. Write down, in terms of $x$, the length of $BC$.     (1)

**(iii)** Write down, in terms of $x$, the perimeter of the triangle. (2)

**(iv)** Factorise your expression for the perimeter of triangle *ABC*. (1)

The perimeter of triangle *ABC* is 30 cm.

**(v)** Use your answer to part (iii) to find the value of $x$. (3)

27 Marie is five years older than Julia, who is $n$ years old.

**(i)** How old, in terms of $n$, is Marie? (1)

Four years ago, Marie was twice as old as Julia was.

**(ii)** Form an equation and solve it to find the value of $n$. (2)

**(iii)** How old will Marie be in three years' time? (2)

28 Buns cost $b$ pence each.

**(i)** What is the total cost of eight buns? (1)

**(ii)** I have $z$ pence to spend. How much change do I receive when I buy 8 buns? (1)

**(iii)** What is the cost of $x$ buns? (1)

Crumpets cost $c$ pence each.

**(iv)** If Joe buys $x$ buns and $y$ crumpets, how much change does he receive from £5? (2)

29 Peter, Quentin and Rachel are playing marbles.

At the start, Quentin has four more marbles than Peter and Rachel has two fewer marbles than Peter.

They play three games in pairs, one person sitting out each time. The results of the games are recorded in this table.

| | Peter | Quentin | Rachel |
|---|---|---|---|
| At the start | $x$ | | $x - 2$ |
| Game 1: Peter loses 2 to Quentin | | $x + 6$ | $x - 2$ |
| Game 2: Peter loses 4 to Rachel | | $x + 6$ | |
| Game 3: Quentin loses ... to Rachel | $x - 6$ | | $x + 3$ |

When the game begins, Peter has $x$ marbles.

**(i)** Copy and complete the table by entering expressions for the number of marbles each person has at each stage. (7)

When all three games are over, Rachel has four times as many marbles as Peter.

**(ii)** Write an equation in $x$ and solve it to find $x$. (3)

(iii) Use your value of $x$ to find how many marbles they have altogether, and how many marbles each person has, when all three games are over. (2)

**30** Solve the following equations:

(i) $a + 4 = 12$ (1)      (iii) $2c = 12$ (1)

(ii) $b - 4 = 7$ (1)      (iv) $2d + 5 = 13$ (2)

**31** Solve the following equations:

(i) $x + 3 = 5$ (1)      (iii) $4z = 6$ (2)

(ii) $y + 5 = 3$ (2)

**32** Solve the following equations:

(i) $2a + 5 = 13$ (2)

(ii) $\frac{1}{4}b = 6$ (1)

(iii) $17 - 2c = 9 + 2c$ (2)

(iv) $6 - d = 2$ (1)

**33** Solve the following equations:

(i) $a - 5 = 13$ (1)

(ii) $13 = 7 + 3b$ (2)

(iii) $4c + 7 = 25 - 2c$ (3)

(iv) $3(d - 4) = 1$ (2)

**34** Solve:

(i) $3(x + 2) = 9$ (2)

(ii) $3(y + 2) = 4(2 - y)$ (3)

(iii) $4 - 2(3z + 5) = 0$ (3)

**35** Solve the following equations:

(i) $x + 4 = 3$ (1)      (iii) $3z - 7 = 11$ (2)

(ii) $6y = 15$ (2)

**36** Solve the following equations:

(i) $5y - 4 = 3y + 3$ (1)

(ii) $5 - x = 7$ (1)

Level 2
Level 2
Level 2
Level 2
Level 2
Level 3
Level 2

(iii) $\frac{3}{4}(z + 1) = 6$ (2)

37 Solve the following equations:

    (i)    $4(r - 3) = r + 15$ (2)

(ii) $7 - 3q = 10$ (2)

(iii) $\frac{4(s+2)}{3} = 3$ (3)

38 Solve the following equations:

    (i)    $3a - 7 = 5$ (2)

(ii) $8 + \frac{1}{3}b = 2$ (2)

(iii) $4(c - 5) = 3c + 5$ (2)

(iv) $\frac{3}{5}d = 30$ (2)

39 Solve the equations:

    (i)    $7u = 56$ (2)

    (ii)   $3v - 5 = v + 9$ (2)

    (iii)  $4(w - 3) = 3w$ (3)

(iv) $\frac{x}{4} + 9 = 3$ (3)

(v) $\frac{3y}{4} = \frac{1}{3}$ (3)

40 (a) Solve the equation: $\frac{1}{4}x + 5 = 13$ (2)

(b) List the negative integers (whole numbers) which satisfy the inequality
$3x + 8 > {}^-7$ (3)

(c) Solve the following inequality for $x$: $13 - 2x > 10$ (3)

41 (a) Solve the following equations:

    (i)    $\frac{1}{6}x = 12$ (1)

    (ii)   $5 + 3(2y - 1) = 17$ (3)

    (iii)  $0.5z - 1.5 = 0$ (2)

    (b) List the positive integers greater than 10 which satisfy the inequality
$4a - 36 \leq a + 4$ (3)

42 (a) List the positive integers which satisfy: $3x - 2 < 13$ (3)

(b) Solve the inequality: $3(x - 1) < 4x - 8$ (3)

43 (a) (i) Solve the inequality: $5 + 3p < 13$ (2)

(ii) List the positive integers which satisfy the inequality in part (a)(i). (1)

(b) (i) Solve the inequality: $8 - \frac{1}{3}q \geq 3$ (3)

(ii) Write down the largest value $q$ can take in part (b)(i). (1)

44 (i) Solve the following inequalities:

(a) $\frac{1}{4}x + 3 \leq 5$ (2)         (b) $5 - 2x < 9$ (2)

(ii) List the prime numbers which satisfy both inequalities. (1)

45 (a) Solve the equation: $\frac{2}{3}x = 18$ (2)

(b) Solve the inequality: $2(t + 3) < 16$ (3)

(c) Factorise: $2c^2 + 4cd$ (2)

(d) Simplify: $\dfrac{5p^2 \times 3p}{6p^3}$ (2)

46 (a) Solve:

(i) $3(x + 5) = 12$ (2)

(ii) $\frac{1}{4}x + 3 = x - 3$ (3)

(b) Simplify:

(i) $6x^2 \times 2xy^2$ (3)       (iii) $\dfrac{(3x)^2 - 5x^2}{2x}$ (3)

(ii) $\dfrac{8x - 6x}{2x}$ (2)

## Extension
Questions 47–51 are included for interest.

47 Showing all your working, solve, by trial and improvement, the equation $x(x + 5) = 100$, giving your answer to 1 decimal place. Copy and complete the table. A start has been made for you. (4)

| $x$ | $x + 5$ | $x(x + 5)$ |
|---|---|---|
| 8 | 13 | 104 |
| 7 | 12 | 84 |
|  |  |  |
|  |  |  |
|  |  |  |
|  |  |  |
|  |  |  |

**48** For the function $y = x^2 - 3x$ find, by trial and improvement, the value of $x$ when $y$ is 20

Give your answers to 2 significant figures. (5)

**49** Solve, by trial and improvement, the equation $x^2 + 4x - 7 = 0$, giving your answer to 2 decimal places. (5)

**50** A rectangle of width $w$ is 1.2 cm longer than it is wide.

   **(i)** Write an expression in terms of $w$ for

      **(a)** the length of the rectangle (1)

      **(b)** the area of the rectangle. (2)

   The rectangle has area 243 cm²

   **(ii)** Find, showing clearly how you do this,

      **(a)** the width of the rectangle (4)

      **(b)** the perimeter of the rectangle. (2)

**51** Sam has thought of a number which he calls $s$.

   Rosie has thought of a number which is smaller than Sam's number.

   The difference between their numbers is 3.5

   **(i)** Write an expression in terms of $s$ for

      **(a)** Rosie's number (1)

      **(b)** the product of their numbers. (2)

   The product of the two numbers is 36

   **(ii)** Find, showing clearly how you do this, the two numbers. (3)

# 4.2 Sequences and functions

In this section the questions cover the following topics:

- Sequences
- Linear functions and their graphs
- Quadratic functions and their graphs
- Simultaneous equations

Many questions cover several topics.

*In this section questions should be answered without using a calculator.*

Level 3

Level 3

1 (a) Write down the next two terms in each of the following sequences:

    (i)   47, 43, 39, 35, ... (2)

    (ii)  1, 3, 7, 13, ... (2)

    (iii) 800, 400, 200, 100, ... (2)

  (b) A sequence starts with 1

    Using the rule 'multiply by 2 and then add 4', write down the next two terms of the sequence. (2)

2 Write down the next two terms in each of the following sequences:

    (i) 32, 27, 22, 17, ...  (1)        (iii) 1, 3, 6, 10, ... (2)

    (ii) 1, 2, 4, 8, ...  (2)        (iv) $1, \frac{11}{12}, \frac{5}{6}, \frac{3}{4}, \frac{2}{3}, \dots$ (2)

3 (a) Write down the next two terms in each of the following sequences:

    (i)   $\frac{1}{20}, 0.1, \frac{3}{20}, 0.2, \dots$ (2)

    (ii)  0, 2, 6, 14, 30, ... (2)

    (iii) 1, 8, 27, 64, ... (2)

  (b) Give two possible answers, each with a reason, for the next term in the sequence: 1, 2, 3, ... (2)

4 Write down the next two numbers in each of the following sequences:

    (i) 2, 9, 16, 23, ...  (2)        (iii) $^{-}12, ^{-}4, 0, 2, \dots$ (2)

    (ii) 1, 3, 9, 27, ...  (2)

5 (a) Write down the next two terms in each of the following sequences:

    (i)   $^{-}5, ^{-}2, 1, 4, \dots$ (2)

    (ii)  11, 21, 41, 71, ... (2)

    (iii) 0.005, 0.05, 0.5, ... (2)

  (b) Apply the rule 'square then add 2' to write down the 2nd, 3rd and 4th terms of the sequence beginning with 1 (2)

6 (a) (i) Write down the next three terms in the sequence:

1, 1.25, 1.5, 1.75, 2, ... (2)

(ii) Find the sum of the first eight terms of the sequence in part (i). (2)

(iii) How many more terms would be needed to make a total of 28.5? (2)

(b) Consider the formula $\frac{1}{2}(n + 1)$ for finding the $n$th term of a sequence.
What is:

(i) the first term (2)

(ii) the tenth term (2)

(iii) the 20th term (2)

(iv) the 99th term? (2)

7 Consider the sequence 1, 5, 9, 13, ...

(i) Write down the next two terms in the sequence. (1)

(ii) Write down the 10th term in the sequence. (1)

Level 3 ■ (iii) Write down the $n$th term in the sequence. (2)

Level 3 ■ (iv) Find the value of $n$ when the $n$th term equals 101 (2)

8 The sequences A and B are:

sequence A: 1, 2, 3, 4, ...

sequence B: 2, 4, 6, 8, ...

Each term of the sequence C is obtained by multiplying together the corresponding terms of A and B, so $C_1 = A_1 \times B_1$, $C_2 = A_2 \times B_2$ and so on.

sequence C: 2, 8, 18, 32, ...

(i) Calculate the next four terms of sequence C. (4)

Level 2 ● (ii) If $C_n = 800$, find the value of $n$. (3)

Level 3 ■ (iii) Write down formulae for the $n$th terms of sequences
A, B and C. (1) (1) (2)

Level 2 ● 9 (i) (a) Find the first four terms, $t_1$, $t_2$, $t_3$, and $t_4$ of the sequence $t_n = n^2 - n$ (2)

(b) Find the 20th term, $t_{20}$ (1)

(ii) Consider the sequence 0, 2, 6, 12, 20, …

   (a) Show how this sequence could be generated as follows:

   $t_1 = 0, \quad t_2 = 0 + 2, \quad t_3 = 0 + 2 + 4, \quad t_4 = 0 + 2 + 4 + 6$ and so on by calculating $t_5$                                                       (1)

   (b) Use your answers to part (i) to find the sum of the numbers

   $0 + 2 + 4 + 6 + 8 + \cdots + 20$                                                (2)

(iii) The sequence 0, 2, 6, 12, 20, … can be represented in the following two ways as patterns of dots.

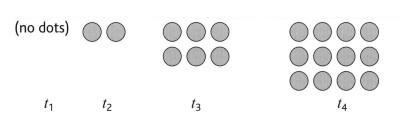

(no dots)

(no dots)

$t_1$               $t_2$                 $t_3$                          $t_4$

   Draw the dot patterns for $t_5$                                                 (2)

10  On a copy of the grid, draw and label the lines with equations:

   (i)   $x = 5$           (1)                                       (iii) $y = x$    (2)

   (ii)  $y = {}^-3$       (1)

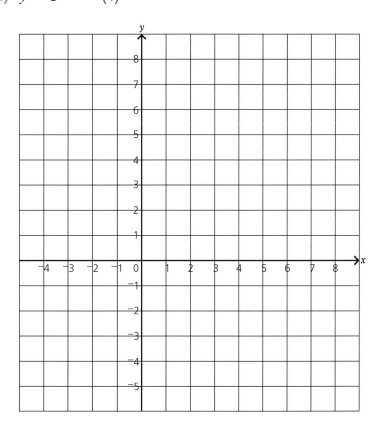

**11 (i)** Study the two function machines A and B.

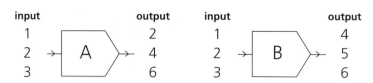

(a) What does machine A do? (1)

(b) What does machine B do? (1)

**(ii)** The machines are put together, as shown here.

input → A → B → output

(a) What is the output if the input is 7? (1)

(b) What is the input if the output is 23? (2)

**(iii)** The order of the machines is changed, as shown here.

input → B → A → output

(a) What is the output if the input is 7? (1)

(b) What is the input if the output is 23? (2)

**12** Consider this two-stage function machine.

input
6
5
→ × 3 → ? → output
27
24

**(i)** What does the second stage of the function machine do? (1)

**(ii)** Fill in the blanks in the table. (2)

| Input | Output |
|-------|--------|
| 4     |        |
|       | 42     |

**13 (i)** Given that $y = x - 1$, copy and complete this table of values. (2)

| $x$ | 0 | 2 | 4 |
|-----|---|---|---|
| $y$ |   |   |   |

**(ii)** Draw the grid and add the line $y = x - 1$ (2)

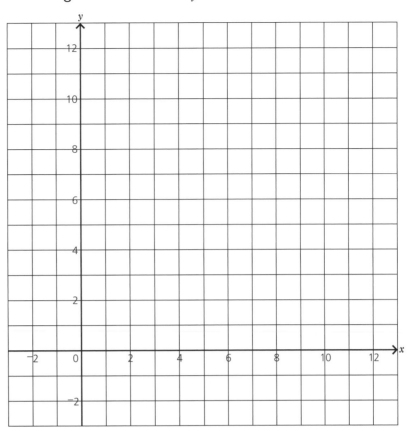

**(iii)** By completing a table of values like that given in part (i), draw the line $y = 10 - x$ on your grid. (3)

| $x$ | | | |
|---|---|---|---|
| $y$ | | | |

**(iv)** Write down the co-ordinates of the point where $y = x - 1$ and $y = 10 - x$ intersect. (1)

**14** This function machine multiplies by 3 and then adds 1

input                  output

$x \rightarrow$ × 3 $\rightarrow$ + 1 $\rightarrow y$

**(i)** Copy and complete this table of input and output values. (3)

| Input | Output |
|---|---|
| 1 | |
| 2 | |
| | 13 |

**(ii)** Complete the equation which represents this function: $y =$ (2)

**Level 2** ●

**15** This function machine squares the input and then adds 1

input                  output

$x \rightarrow$ square $\rightarrow$ + 1 $\rightarrow y$

**(i)** Copy and complete this table of input and output values.

| Input | Output |
|-------|--------|
| 0 | |
| 1 | |
| 2 | |
| 3 | |
| | 50 |

(1)
(1)
(1)
(1)
(2)

**(ii)** Complete the equation which represents this function: $y =$ (2)

**Level 3**

**16 (i)** Calculate the values of $y$ to complete the table for the function $y = x^2 - 2$ (3)

| $x$ | $^-2$ | $^-1$ | 0 | 1 | 2 | 3 |
|-----|-------|-------|---|---|---|---|
| $y$ | 2 | | | $^-1$ | 2 | |

**(ii)** Draw the grid, plot the corresponding points and draw the curve $y = x^2 - 2$ (3)

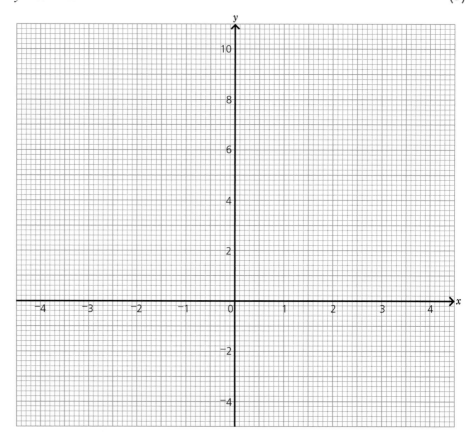

**(iii)** Copy and complete the table for the straight line $y = 2x + 1$, then plot the corresponding points and draw the straight line graph on your grid. (3)

| $x$ | $^-2$ | 0 | 2 |
|-----|-------|---|---|
| $y$ | | 1 | |

**(iv)** Write down the co-ordinates of the points of intersection of the curve and the straight line. (2)

Level 3

**17** This table of values has been prepared for a function.

| x | ⁻3 | ⁻2 | ⁻1 | 0 | 1 | 2 | 3 |
|---|---|---|---|---|---|---|---|
| y | 13 | 8 | 5 | 4 | 5 | 8 | 13 |

(i) Complete the equation for this function: $y = $ ... (2)

(ii) Complete the ordered pair $(x, y)$ which satisfies the function:

$$\left(\frac{1}{2}, \cdots\right)$$ (2)

Level 3

**18** Here is a table of values for $y = 2x^2 - 3$

| x | ⁻3 | ⁻2 | ⁻1 | 0 | 1 | 2 | 3 |
|---|---|---|---|---|---|---|---|
| y | 15 | 5 | ⁻1 | ⁻3 | ⁻1 | 5 | 15 |

(i) Using the table of values, draw the graph of $y = 2x^2 - 3$ on a copy of the grid. (2)

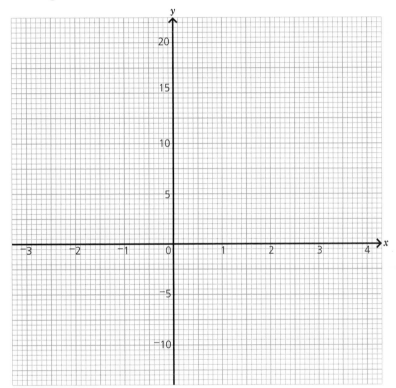

(ii) Given that $y = x + 3$ copy and complete the table of values. (2)

| x | ⁻2 | 0 | 2 | 4 |
|---|---|---|---|---|
| y | | | 5 | |

(iii) On your grid, draw the graph of $y = x + 3$ (1)

(iv) Write down the co-ordinates of the points where the graphs of $y = 2x^2 - 3$ and $y = x + 3$ intersect. (2)

19 By copying and completing these tables of values, find two points of intersection of the quadratic function $y = \dfrac{x^2}{4}$ and the linear function $y = x$ (4)

$y = \dfrac{x^2}{4}$

| $x$ | $^-2$ | $^-1$ | 0 | 1 | 2 | 3 | 4 |
|---|---|---|---|---|---|---|---|
| $y$ | | | | | | | |

$y = x$

| $x$ | $^-2$ | $^-1$ | 0 | 1 | 2 | 3 | 4 |
|---|---|---|---|---|---|---|---|
| $y$ | | | | | | | |

20 (i) Copy and complete the table of values for the function $y = 4 - \dfrac{1}{2}x^2$ (5)

| $x$ | $^-3$ | $^-2$ | $^-1$ | 0 | 1 | 2 | 3 | 4 |
|---|---|---|---|---|---|---|---|---|
| $y$ | $^-0.5$ | | | | 3.5 | | | $^-4$ |

(ii) Copy the grid and draw the graph of $y = 4 - \dfrac{1}{2}x^2$ (3)

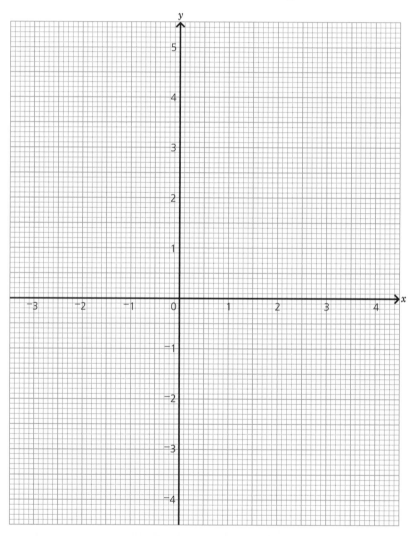

(iii) Copy and complete the table of values for the linear function $y = 1 - x$ (1)

| $x$ | $^-3$ | 0 | 3 |
|---|---|---|---|
| $y$ | | | $^-2$ |

(iv) Draw the line $y = 1 - x$ on your grid. (2)

(v) Write down the co-ordinates of the points of intersection of the curve and the line. (2)

**Level 3**

**21 (i)** Copy and complete the table of values for the function $y = x^2 - 4$ (2)

| $x$ | ⁻2 | ⁻1 | 0 | 1 | 2 | 3 | 4 |
|-----|-----|-----|-----|-----|-----|-----|-----|
| $y$ | 0 |  | ⁻4 |  | 0 |  |  |

**(ii)** Copy the grid, plot the points and draw the graph of the function $y = x^2 - 4$ within the limits of ⁻2 ≤ $x$ ≤ 4 (3)

$y = x^2 - 4$

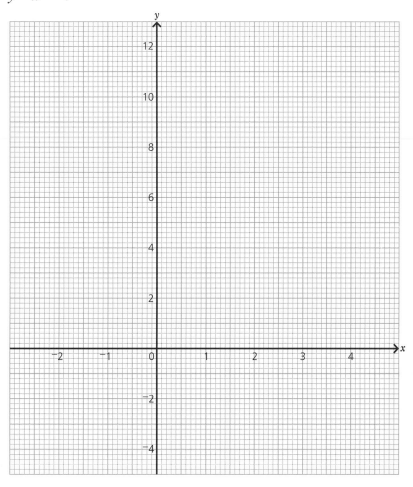

**(iii)** Copy and complete the table when $y = 3x - 1$ (2)

| $x$ | ⁻1 | 0 | 2 |
|-----|-----|-----|-----|
| $y$ |  |  | 5 |

**(iv)** Draw the graph of the line $y = 3x - 1$ on your grid. (2)

**(v)** State the co-ordinates of the points of intersection of the two graphs. (2)

**22 (i)** By substituting values for $x$ into the equation $y = x^2 - 3$, complete your table of values for $y$. (3)

| $x$ | $^-3$ | $^-2$ | $^-1$ | 0 | 1 | 2 | 3 |
|-----|-----|-----|-----|-----|-----|-----|-----|
| $y$ | | 1 | | $^-3$ | | | |

**(ii)** Draw the graph of $y = x^2 - 3$ (3)

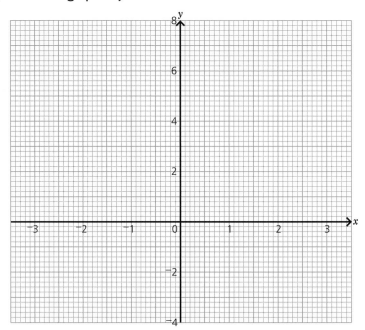

**(iii)** Substitute values of $x$ into the equation $y = 4 - x$ to copy and complete the table of values for $y$. (2)

| $x$ | $^-3$ | 0 | 2 |
|-----|-----|-----|-----|
| $y$ | | | 2 |

**(iv)** On your grid draw the graph of $y = 4 - x$ (1)

**(v)** Write down the co-ordinates of the points of intersection of the two graphs. (2)

**23 (i) (a)** Copy and complete the table for the function $y = x^2 + 3$ (2)

| $x$ | $^-2$ | $^-1$ | 0 | 1 | 2 |
|-----|-----|-----|-----|-----|-----|
| $y$ | | 4 | | | |

**(b)** Copy the grid and draw the graph of $y = x^2 + 3$      (3)

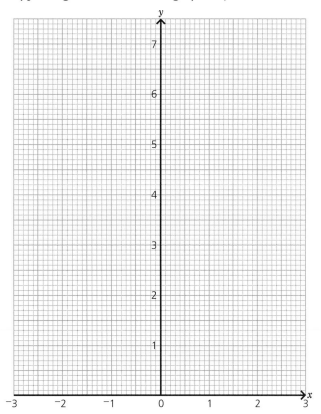

**(ii) (a)** Copy and complete the table for the function $y = 5 - \frac{1}{2}x$      (2)

| $x$ | ⁻2 | 0 | 2 |
|---|---|---|---|
| $y$ | | | |

**(b)** Draw the graph of $y = 5 - \frac{1}{2}x$ on your grid.      (2)

**(iii)** Estimate the co-ordinates of the point of intersection of the two graphs in the first quadrant.

Give your answer correct to 1 decimal place.      (2)

**Level 3**

**24 (i)** Given that $y = 2x^2 - 4$, copy and complete this table of values.      (3)

| $x$ | ⁻3 | ⁻2 | ⁻1 | 0 | 1 | 2 | 3 |
|---|---|---|---|---|---|---|---|
| $y$ | | 4 | | | | | 14 |

**(ii)** Copy the grid and draw the graph of $y = 2x^2 - 4$      (2)

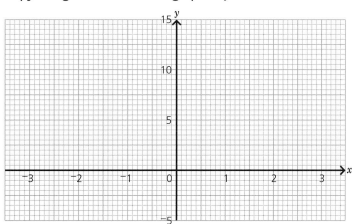

**(iii)** Given that $y = 8 - x$, copy and complete this table of values. (2)

| $x$ | $^-3$ | 0 | 3 |
|---|---|---|---|
| $y$ | | | |

**(iv)** On your grid, draw the graph of $y = 8 - x$ (1)

**(v)** Write down the co-ordinates of the points where the graphs of $y = 2x^2 - 4$ and $y = 8 - x$ intersect. (2)

Level 3

**25 (i)** Copy and complete this table of values for the function $y = 4 - x^2$ (3)

| $x$ | $^-3$ | $^-2$ | $^-1$ | 0 | 1 | 2 | 3 |
|---|---|---|---|---|---|---|---|
| $y$ | $^-5$ | | | 4 | 3 | | |

**(ii)** Copy the grid and draw the graph of $y = 4 - x^2$ (2)

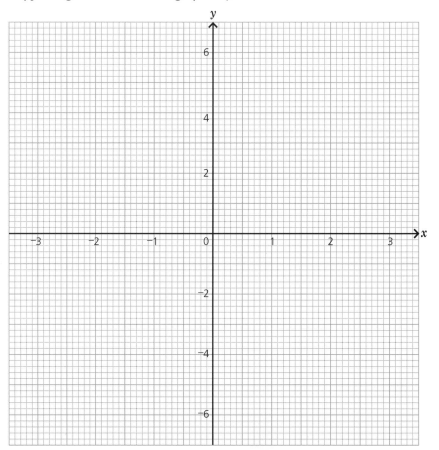

**(iii)** Copy and complete the table for the linear function $y = 2x + 2$ (2)

| $x$ | $^-3$ | $^-1$ | 1 |
|---|---|---|---|
| $y$ | | | |

**(iv)** Draw the graph of the line $y = 2x + 2$ on your grid. (1)

**(v)** Estimate the co-ordinates of the points of intersection of the two graphs. (2)

Level 3

**26** Oranges cost $x$ pence each and lemons cost $y$ pence each.

**(i)** 4 oranges and 3 lemons cost £1.70

Write this information as an equation in terms of $x$ and $y$. (1)

(ii) 3 oranges cost the same as 2 lemons.

Write this information as an equation in terms of $x$ and $y$. (1)

(iii) Use your answers to parts (i) and (ii) to find the values of $x$ and $y$. (4)

(iv) Felicity spends exactly £2 when buying a selection of oranges and lemons.

Write down all the different combinations of oranges and lemons which she could buy. (3)

Level 3 27 $x$ and $y$ are two positive whole numbers.

$x$ is three and a half times as large as $y$.

(i) Write down an equation, in terms of $x$ and $y$, to show this. (1)

$x$ is two more than 3 times $y$.

(ii) Write down an equation, in terms of $x$ and $y$, to show this. (1)

(iii) Using your answers to parts (i) and (ii), solve the equations to find the values of $x$ and $y$. (4)

(iv) Using your answers to part (iii), write the ratio of $\sqrt{14x}$ to $\sqrt{y}$ in its lowest terms. (2)

Level 3 28 Solve the simultaneous equations

$3a - 5b = 49$

$a + 2b = 31$ (4)

Level 3 29 Tom adds two numbers and then multiplies the result by 5

He gets 80

He says that the difference between his two numbers is 7

(i) Write two equations to represent this information. (3)

(ii) Solve the two equations simultaneously to find Tom's two numbers. (3)

Level 3 30 A rectangle has length $l$ and width $w$.

(i) Write an expression for

(a) the difference between the length and the width (1)

(b) the perimeter of the rectangle. (2)

The rectangle is 6 cm longer than it is wide. The perimeter of the rectangle is 42 cm.

(ii) Write down and solve the two equations simultaneously to find the length and width of the rectangle. (3)

(iii) Calculate the area of the rectangle. (1)

# Geometry and measures

## 5.1 Measures

In this section the questions cover the following topics:

- Metric units and imperial units
- Constructions
- Areas and volumes
- Circles
- Speed

Many questions cover several topics.

*In this section you should answer the questions without using a calculator except where indicated by* 🖩

1   When answering this question, use the following conversion rates:

> **1 pound = 0.45 kilogram**
>
> **14 pounds = 1 stone**

   **(i)**  A box of *Scrummy* dog biscuits has a mass of 900 grams.

       What is this mass in pounds? (2)

   **(ii)**  Alan's mass is 7 stone 2 pounds.

       Write Alan's mass in kilograms. (3)

2   Use the conversion 1 inch = 2.54 cm in this question.

   **(i)**  A piece of string is 50 inches long. Express this length in metres. (2)

   **(ii)**  A door is 2.0 metres high. Express the height in inches to the nearest inch. (2)

3   In this question, use the conversion

> **1 hectare = 2.47 acres**
>
> **1 hectare = 10 000 m²**

   **(a)**  A woodland has an area of 200 000 m².

       What is this area in

       **(i)**  hectares (1)

       **(ii)**  acres? (2)

**(b)** A building plot is advertised as having an area of $\frac{1}{4}$ acre.

What is the area (to the nearest 100 m²) of the plot in square metres? (3)

4 Plastic washing up bowls are made in various shapes, but many are rectangular, with rounded corners and sloping sides.

**(i)** Estimate the volume of water which an ordinary plastic washing up bowl would hold, by copying and completing the following table. (3)

|        | Estimate |
|--------|----------|
| Length | cm |
| Width  | cm |
| Depth  | cm |
| Volume | cm³ |

**(ii)** Give your answer to part (i) in litres to 1 significant figure. (2)

5 **(i)** Choose an estimate of the mass of an ordinary tennis ball from the following list. (2)

| 10 g | 60 g | 120 g | 250 g | 500 g |

**(ii)** Choose an estimate of the circumference of an ordinary tennis ball from the following list. (2)

| 5 cm | 10 cm | 20 cm | 40 cm | 50 cm |

**(iii)** Choose an estimate of the volume of an ordinary tennis ball from the following list. (2)

| 5 cm³ | 30 cm³ | 110 cm³ | 220 cm³ | 500 cm³ |

6 **(i)** On your paper measure a space of 15 cm. Mark a point *A* with a cross, 2 cm from the bottom of the space and half way across the page. Construct triangle *ABC* given that *AB* = 8 cm, angle *ABC* = 40° and angle *BAC* = 110°. (3)

**(ii)** By taking suitable measurements, calculate the perimeter of triangle *ABC*. (2)

7  (i)   On your paper, construct triangle ABC with AB = 10.5 cm, BC = 6.5 cm
         and angle ABC = 55°.                                                      (3)

   (ii)  Measure and write down the shortest distance from C to AB.                (2)

   (iii) Measure and write down the size of the angle BCA.                         (2)

8  The distance from A to B is 10.8 metres.

   (a) (i)   In the middle of a 15 cm space on your page, use a scale of
             1 : 100 to make a scale drawing of the line AB.                       (2)

       (ii)  Construct the perpendicular bisector of AB.                          (2)

   (b) A square has diagonals of length 10 centimetres.

       (i)   On your paper, make an accurate drawing of the square.                (3)

       (ii)  By taking suitable measurements, find the perimeter of the square.    (2)

9  In quadrilateral ABCD, angle BAD = 90°, AB = AD = 5 cm and BC = DC = 7 cm.

   (i)   Starting in a space on your page, as shown here, make an accurate
         drawing of quadrilateral ABCD.                                            (4)

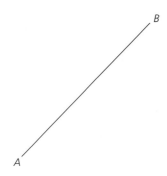

   (ii)  Draw any lines of symmetry on your quadrilateral.                         (1)

   (iii) Which type of quadrilateral have you drawn?                               (1)

   (iv)  Construct the perpendicular to AC passing through D.                      (2)

   (v)   Comment on your observations.                                             (1)

68

10 The line *AC* is a diagonal of the rhombus *ABCD*. The length of the side *AB* is 6 cm. The length of *AC* is exactly 10 cm and this diagram shows how this line would appear in a 12 cm space.

*A* ——————————————————————————————— *C*

(i) Using pencil, ruler and compasses, construct accurately the rhombus *ABCD*. (3)

(ii) Measure the length, in cm, of the diagonal *BD*. (1)

(iii) Calculate the area, in cm², of the rhombus *ABCD*. (2)

(iv) Construct the bisector of angle *BAC*. (2)

11 The angles in quadrilateral *ABCD* taken in order are in the ratio 1 : 3 : 5 : 3

(i) Calculate the angles of the quadrilateral. (2)

(ii) If *AB* and *AD* are each 9 cm, make an accurate construction of the quadrilateral *ABCD*. Start in a space with *AB* positioned as shown here. (3)

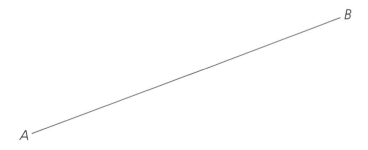

(iii) By taking suitable measurements, find the area of the quadrilateral. (3)

**12 (i)** What is the shape drawn here? (1)

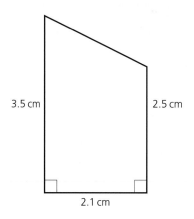

3.5 cm

2.5 cm

2.1 cm

**(ii)** Calculate the area of the shape. (4)

**13** Mohammed wishes to make a box to contain 48 one centimetre cubes.

One cuboid, A, which he could make to contain the 48 cubes, is drawn on this isometric dotted grid.

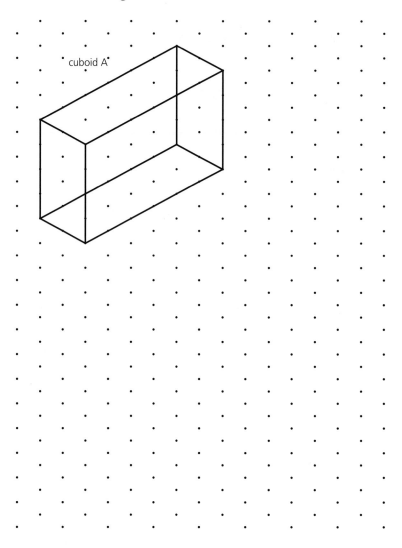

cuboid A

**(i)** On a similar grid draw a different shaped cuboid which Mohammed could make to contain the same 48 cubes. Do not worry if some lines overlap in your drawing.

Label your drawing cuboid B. (3)

**(ii)** The total surface area of a cuboid is found by adding together the areas of the six faces of the cuboid.

Work out the total surface area, in cm², of cuboid A. (4)

**14** A cake is in the shape of a cuboid.

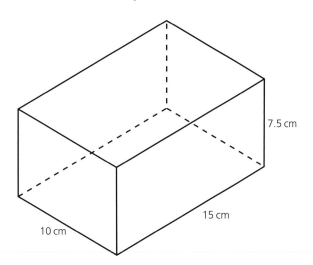

The cake is 15 cm long, 10 cm wide and 7.5 cm high.

**(i)** Calculate the volume, in cm³, of the cake. (4)

All four sides and the top of the cake are to be covered in icing.

**(ii)** Calculate the area which is to be covered in icing. (4)

**15** A block of wood is a cuboid with length 9 cm, width 5 cm and height 4 cm.

**(i)** On an isometric dotted grid, copy and complete the drawing of the cuboid.

Position one complete edge as shown in this diagram. (2)

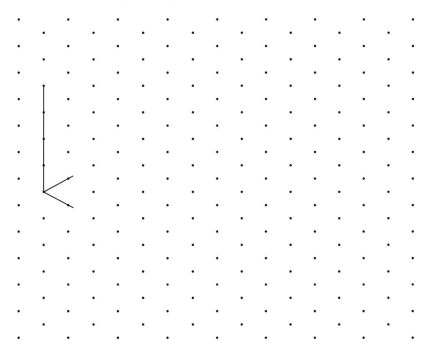

**(ii)** Calculate the volume of the block. (1)

A square hole measuring 1 cm by 1 cm by 4 cm is cut down through the middle of the top face of the block.

**(iii)** Draw the shape of the hole on your cuboid. (2)

**(iv)** What is the total surface area of the block with the hole through it? (4)

**16** The lengths of the edges of two cubes are in the ratio 1:4

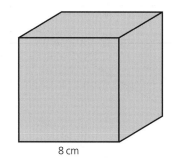

2 cm            8 cm

**(i)** Write down the ratio of the surface areas of the two cubes. (2)

**(ii)** Write down the ratio of the volumes of the two cubes. (2)

**17** The scale drawing here shows a square field.

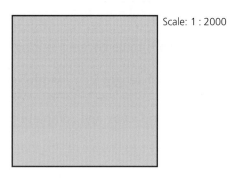

Scale: 1 : 2000

**(i)** What is the perimeter, in metres, of the field? (2)

**(ii)** What is the area, in hectares, of the field? (2)

**18** A circular table top has a radius of 55 cm.

55 cm

**(i)** Find the area of the table top, in square metres, to 3 significant figures. (2)

Another table has half the radius of the first table.

**(ii)** Use your answer to part (i) to write down the area of the top of the smaller table, in square metres, to 3 significant figures. (2)

A cafe has 8 of the large tables and 12 of the small tables. The tops of all of these tables are to be varnished. One tin of varnish will cover an area of 4 m². 

**(iii)** How many tins of varnish will be needed? (3)

**19** The mean radius, at the equator, of the planet Mars is 3397 km.

**(i)** Use this value to find the circumference, to the nearest kilometre, of Mars. (3)

**(ii)** Write your answer to part (i) correct to the nearest 100 km. (1)

**20 (i)** A circle has diameter 8 cm.

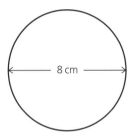

8 cm

**(a)** Calculate the circumference of the circle. (2)

**(b)** Calculate the area of the circle. (3)

**(ii)** A design consists of an equilateral triangle of side 8 cm with a semicircle drawn on each side.

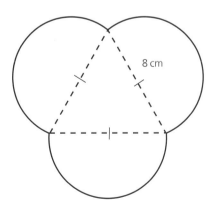

8 cm

**(a)** Calculate the perimeter of the design. (2)

**(b)** Use Pythagoras' theorem to show that the area of an equilateral triangle of side 8 cm is approximately 27.7 cm². (3)

**(c)** Calculate the area of the design to the nearest cm². (Remember calculations are to *3 significant figures*.) (2)

Level 3

**21 (i)** The diameter of the circle is 60 cm.

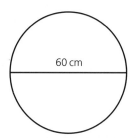

**(a)** Calculate the circumference of the circle. (2)

**(b)** Calculate the area of the circle. (2)

**(ii)** A semi-circular window, diameter 60 cm, is made up of two red sectors of angle 60° and three yellow sectors of angle 20°.

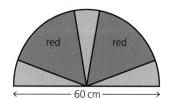

**(a)** Calculate the perimeter of the semi-circular window. (2)

**(b)** Calculate the area of one yellow glass sector. (2)

**(c)** The glass is 6 mm thick. Calculate the volume of one red glass sector. (2)

 **22** My new car tyre has a diameter of 59 cm. (2)

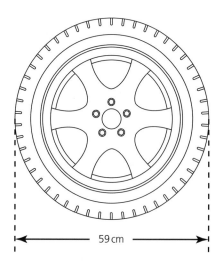

**(i)** Calculate the circumference of the wheel in metres, correct to the nearest centimetre. (2)

**(ii)** Calculate, using your answer to part (i), how many complete revolutions the wheel would make in travelling 100 metres. (2)

A worn tyre makes 56 revolutions in travelling 100 metres.

    (iii)  Calculate the diameter (in cm) of the worn tyre.    (2)

**23 (a)** It takes 45 minutes to travel by train from Glasgow to Edinburgh. Morag catches the 14:28 train from Glasgow. At what time should Morag arrive in Edinburgh?   (2)

  **(b)** A distance runner averages 1 mile every $5\frac{1}{2}$ minutes. How long in hours and minutes will it take him to run 26 miles at this rate?   (2)

  **(c)** Tom cycles at an average speed of 15 km/h.

    **(i)** How long in minutes does it take Tom to cycle 1 kilometre?   (1)

    **(ii)** How far could Tom cycle in 28 minutes?   (1)

  **(d)** **(i)** How many metres are there in 7.2 kilometres?   (1)

    **(ii)** Write 7.2 km/h as a speed in metres per second.   (2)

**24** Janet challenges John to a 50 metre sack race. They begin at the same time. Janet covers the 50 metres in a time of 20 seconds.

  **(i)** What is Janet's average speed, in m/s, for the race?   (2)

John covers the 50 metres at an average speed of 2 m/s.

  **(ii)** How many seconds does John take?   (2)

  **(iii)** By how many seconds does Janet beat John?   (1)

  **(iv)** How far from the finishing line was John when Janet finished?   (2)

## 5.2 Shape

In this section the questions cover the following topics:
- Plane shapes and their properties
- Symmetry
- Solid shapes and their properties
- Nets

Many questions cover several topics.

*In this section questions should be answered without using a calculator.*

1  **(i)**  Copy the grid and plot the following points:

*P* (2, 0)          *Q* (1, ⁻4)          *R* (⁻3, ⁻3)          *S* (⁻2, 1)

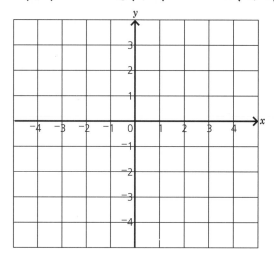

Join the points to form the square *PQRS*.                    (3)

**(ii)**  Draw all lines of symmetry on your square.                    (2)

**(iii)**  What is the order of rotational symmetry of the square *PQRS*?          (1)

2  Sketch, indicating equal sides and lines of symmetry, and name the following:

**(i)**  A quadrilateral with one pair of equal sides and one line of symmetry  (2)

**(ii)**  A quadrilateral with two pairs of equal sides and no lines of symmetry  (2)

3  *ABCDE* is part of a regular octagon.

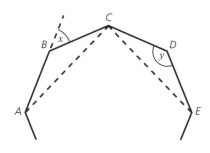

**(i)**  Calculate the size of an exterior angle, marked *x*.                    (2)

**(ii)**  Calculate the size of an interior angle, marked *y*.                    (2)

**(iii)**  Calculate the size of the angle *ACE*.                    (2)

**Level 2** ●

4 *ABCDEF* is part of a regular decagon.

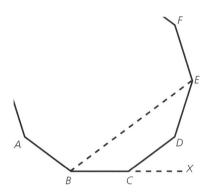

(i) Calculate the size of:

   (a) angle *DCX* (2)

   (b) angle *CDE*. (2)

(ii) What special type of shape is *BCDE*? (1)

(iii) Calculate the size of angle *ABE*. (2)

**Level 3** ▪

5 *PQRST* shows part of a regular 18-sided polygon.

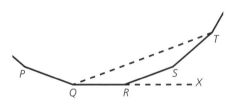

(i) What is the order of rotational symmetry of a regular 18-sided polygon? (1)

*QR* is produced (extended) to *X*.

(ii) Calculate the size of angle *SRX*. (2)

(iii) What is the sum of the interior angles of a regular 18-sided polygon? (2)

(iv) Calculate the size of angle *PQT*. (2)

(v) Which type of figure is *QRST*? (1)

**6** The net of a solid shape is drawn here.

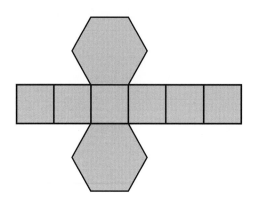

    **(i)**   What is the name of the solid?   (3)

    **(ii)**  How many different ways could the solid fit into a hole which is the same shape and size as the solid?   (2)

**7** A cuboid measures 5 cm by 4 cm by 3 cm.

    **(i)**   On a square dotted grid, draw a net of the cuboid.   (3)

    **(ii)**  Calculate the total surface area of the cuboid in part (i).   (3)

    **(iii)** Calculate the volume of the cuboid in part (i)   (1)

# 5.3 Space

In this section the questions cover the following topics:

- Angles
- Bearings and scale drawings
- Transformations on a grid
- Enlargement

Level 3

- Pythagoras' theorem
- Trigonometry (this section is included for interest only)

Many questions cover several topics.

*In this section questions should be answered without using a calculator except where indicated by* 🖩

1   Find the size of each of the angles marked.

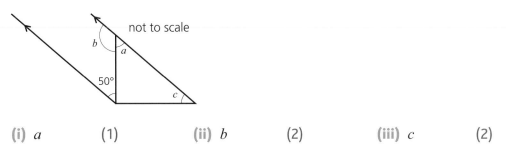

not to scale

(i) *a*          (1)          (ii) *b*          (2)          (iii) *c*          (2)

2   Calculate the angles marked *a*, *b* and *c* on the diagram.

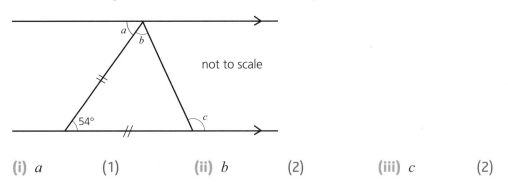

not to scale

(i) *a*          (1)          (ii) *b*          (2)          (iii) *c*          (2)

3   (a) Calculate the sizes of the angles marked *a*, *b* and *c*.

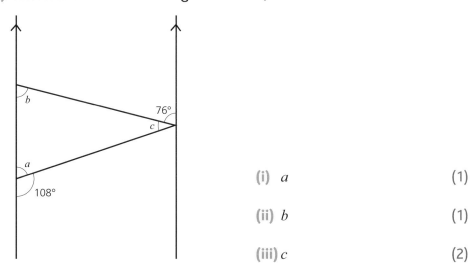

(i) *a*          (1)

(ii) *b*          (1)

(iii) *c*          (2)

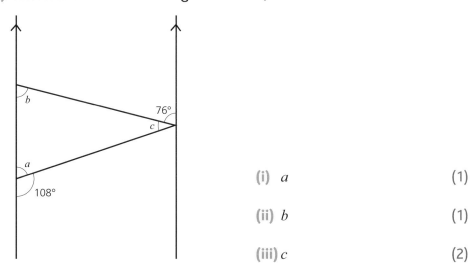

(79)

**(b)** Calculate the angles marked *p*, *q* and *r* on the diagram.

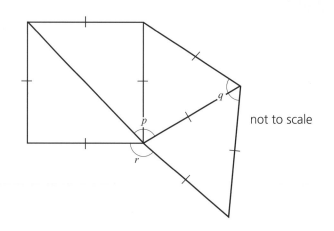

not to scale

**(i)** *p*        (2)        **(ii)** *q*        (2)        **(iii)** *r*        (2)

4  **(a)** Calculate the sizes of the angles marked *a*, *b* and *c*.

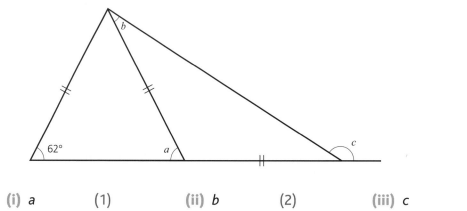

**(i)** *a*        (1)        **(ii)** *b*        (2)        **(iii)** *c*        (2)

**(b)** Calculate the sizes of the angles marked *d* and *e*.

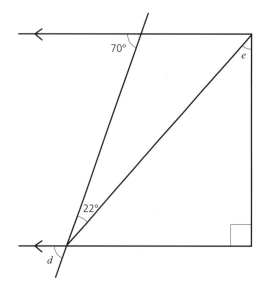

**(i)** *d*                    (1)                    **(ii)** *e*        (2)

5 **(a)** Calculate the size of each of the angles marked *a*, *b* and *c*.

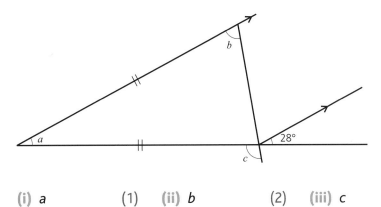

   **(i)** *a*          (1)   **(ii)** *b*          (2)   **(iii)** *c*          (2)

   **(b)** Calculate the size of angle *d*.          (2)

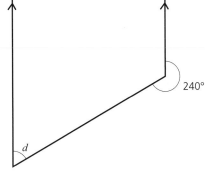

240°

6   The diagram shows a regular pentagon and an equilateral triangle joined to
    form the shape *ABCDEF*.

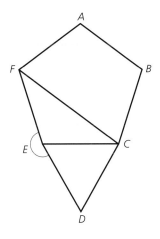

   Find the size of:

   **(i)** angle *ABC*   (2)   **(ii)** angle *ECF*   (2)   **(iii)** reflex angle *DEF*.   (2)

7  **(i)** Showing all your working, calculate the sum of the interior angles of a nonagon (9-sided shape). (3)

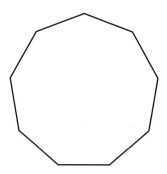

**(ii)** The nonagon *ABCDEFGHI* is symmetrical about the line shown.

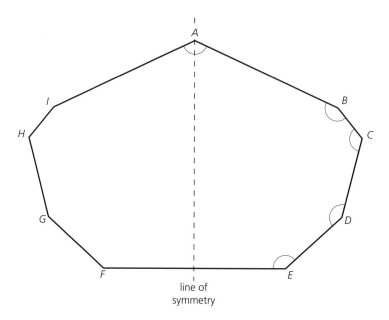

line of symmetry

The angles at vertices *A*, *B*, *C*, *D* and *E* are in the ratio 8:8:6:5:7

**(a)** If the angle at *A* is 8$x$, find the sum of the interior angles of the nonagon in terms of $x$. (2)

**(b)** Calculate the value of $x$. (2)

**(c)** Use your value of $x$ to find the angle at vertex *D*. (1)

8  Doris, Edna, Flora and Gloria are standing on a rounders field.

Doris is 10 m due east of Edna, Flora is 16 m south west of Doris and Gloria is 10 m due north of Doris.

**(i)** Using a scale of 1 cm to represent 2 m, make a scale drawing to show the positions of the four girls. (4)

**(ii)** Use your drawing to find

**(a)** the distance in metres of Flora from Edna (2)

**(b)** the direction Edna must throw the ball to Gloria. (2)

If you are allowed to draw in this book, you will save valuable time when answering the next few questions, as you will not need to copy out the diagrams and grids beforehand.

9   Alfie and Brenda are practising their hockey skills.

The position of Alfie (A) is marked.

Alfie passes the ball 50 metres to Brenda on a bearing of 140°.

(i)   Copy the diagram, leaving space as shown and plot the positions of Alfie (A) and Brenda (B) using a scale of 1 cm : 5 m.    (3)

Alfie now runs 30 m on a bearing of 210°.

(ii)   Plot Alfie's new position (A').    (2)

Brenda (B) now passes the ball back to Alfie (A').

(iii)   Find the distance in metres the ball travels from B to A'.    (1)

(iv)   Find the bearing on which the ball travels from B to A'.    (2)

**10** A coastguard station (C) is positioned on the coast as shown here.

A lifeboat station (L) is positioned on the coast 970 m due east of C.

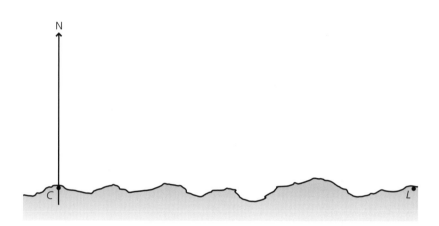

The coastguard receives a distress signal from a trawler (T) which is 1 km from the coastguard station (C) on a bearing of 050°.

(i) Copy the diagram, leaving space on your page as shown. The coastline does not need to be an exact copy, but *remember that L is exactly 970 metres due east of C*. Use a scale of 1:10 000 to plot the position of the trawler (T). (3)

The lifeboat leaves the lifeboat station (L), and sails directly to the trawler.

(ii) Use your scale drawing to find:

(a) the distance in metres that the lifeboat travels to the trawler (2)

(b) the bearing on which the lifeboat travels to get to the trawler. (2)

(iii) If the lifeboat travels at 12 kilometres per hour, how long will it take to reach the trawler? Give your answer to the nearest minute. (2)

11 A and B are two rescue stations 1.08 km apart on opposite sides of a lake and the bearing of B from A is 100°.

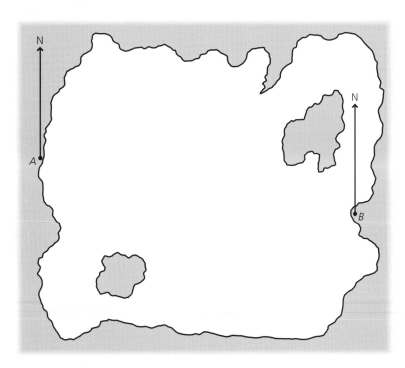

The scale of the map is 1:10000

Late at night, a yacht, Y, is in distress on the lake.

A searchlight at rescue station A picks up the yacht on a bearing of 054°, and a searchlight at rescue station B, picks up the yacht on a bearing of 300°.

(i)   Make an accurate drawing showing the positions of A and B, and then draw two lines on the diagram to locate the position of the yacht, Y.   (3)

*You can use your imagination rather than trace the outline of the lake, but remember that on your drawing the distance AB is 10.8 cm and the bearing of B from A is 100°.*

(ii)   What is the distance, in metres, of the yacht from rescue station A?   (2)

(iii)  What is the bearing of A from B?   (2)

**12** From the base camp (*C*), the bearing of a polar bear (*P*) is 225°.

The weather station (*W*) is 90 m south of *C*. The bearing of the bear from *W* is 285°.

(i)    In a space, as shown here, using a scale of 1:1000, draw a diagram to
       show the positions of the base camp (*C*), the weather station (*W*)
       and the polar bear (*P*).                                              (4)

(ii)   By using measurements on the diagram, find the distance of the polar
       bear from the weather station. Give your answer in metres to the
       nearest metre.                                                         (2)

**13 (i)** On a map with scale 1:100 000, what distance, in kilometres, does
1 centimetre represent? (1)

Bratby (*B*) is 9 km away from Addleton (*A*) on a bearing of 070°.

**(ii)** Copy the diagram and, using a scale of 1:100 000, mark the position
of Bratby. (2)

**(iii)** What is the bearing of Addleton from Bratby? (1)

Creakybridge (*C*) is on a bearing of 115° from Addleton and 210° from Bratby.

**(iv)** Mark the position of Creakybridge on your diagram. (3)

**(v)** What is the distance in kilometres from Bratby to Creakybridge? (2)

**14 (i)** Copy this diagram and then

    **(a)** draw the line $x = 4$                     (1)

    **(b)** reflect triangle **A** in the line $x = 4$ and label the image **B**.     (2)

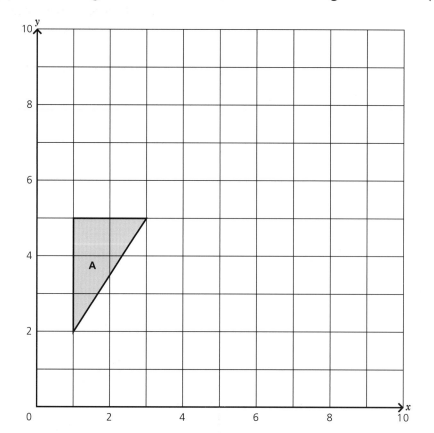

**(ii)** Rotate triangle **A** through 90° clockwise about the point (3, 5) and label the image **C**.     (2)

**(iii)** Translate triangle **A** 6 units to the right and 4 units up. Label the image **D**.     (2)

**(iv)** What is the area of triangle **A**?     (1)

**15 (i)** Copy the grid and plot the points (2, 1), (5, 1) and (3, ⁻1).　　　(1)

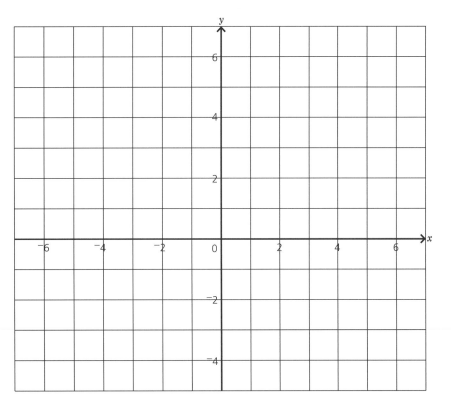

**(ii)** Join the points to form a triangle and label the triangle **A**.　　　(1)

**(iii)** Reflect triangle **A** in the $y$-axis. Label the image **B**.　　　(2)

**(iv)** Rotate triangle **A** through 180° about the point (0, 2). Label the image **C**.　　　(2)

**(v)** Describe the single transformation that would transform **B** onto **C**.　　　(2)

**(vi)** Draw a triangle **D** so that all four shapes form a pattern with rotational symmetry of order 2　　　(2)

**16 (i)** Copy the grid and plot the points (⁻3, 1), (⁻2, 5) and (⁻1, 2). Join the points and label the triangle **A**. (2)

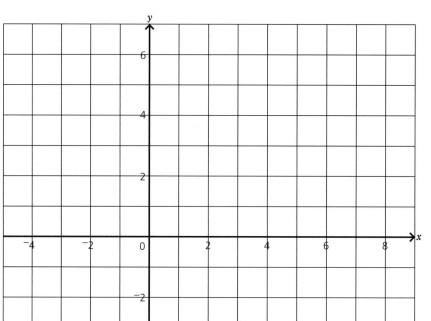

**(ii) (a)** Draw and label the line $x = 2$ (1)

**(b)** Reflect triangle **A** in the line $x = 2$ and label the image **B**. (2)

**(iii)** Rotate triangle **A** through 90° anticlockwise about the point (0, 2). Label the image **C**. (3)

**(iv)** Draw a fourth triangle so that the pattern, made by the four triangles, has one line of symmetry. Label the fourth triangle **D**. (1)

**17 (i)** Copy the grid and plot the points (1, 1), (3, 1) and (1, ⁻2). (1)

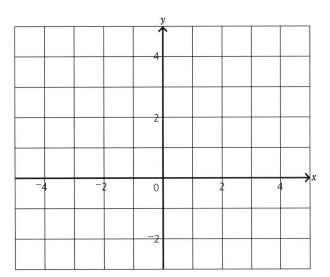

(ii) Join the points to form a triangle and label the triangle **A**. (1)

(iii) Reflect triangle **A** in the *y*-axis. Label the image **B**. (2)

(iv) Rotate triangle **A** through 90° anticlockwise about the point (0, 0).

Label the image **C**. (2)

(v) A reflection maps **B** onto **C**. On the grid, draw the mirror line. (1)

(vi) Give the equation of the mirror line drawn in part (v). (1)

**18** Copy the drawing. With centre $O$, copy the diagram and enlarge the shape by scale factor 2 (2)

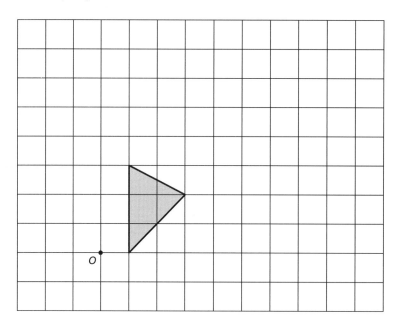

**19 (i)** Copy the diagram. Taking $P$ as the centre of enlargement, enlarge triangle **A** using a scale factor of 3

Label the image **B**. (3)

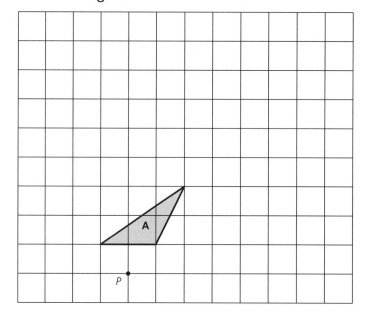

**(ii)** Find the area of triangle **A**. (2)

**(iii)** Hence, or otherwise, find the area of the enlarged triangle, **B**. (2)

**20 (i)** On a copy of this grid, plot and label the points $A(0, 1)$, $B(2, 1)$, $C(3, 3)$ and $D(1, 3)$. (2)

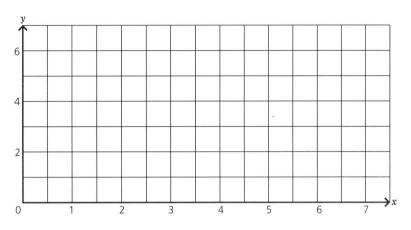

**(ii)** Join the points $A$, $B$, $C$ and $D$ to form a parallelogram. (1)

**(iii)** Enlarge $ABCD$ using the origin $(0, 0)$ as the centre and with scale factor of 2

Label your enlargement $A'B'C'D'$ and mark the centre of the parallelogram $A'B'C'D'$ as $P$. (2)

**(iv)** Write down the co-ordinates of $P$. (1)

**(v)** What is the ratio of the area of $A'B'C'D'$ to the area of $ABCD$? (2)

**21 (i)** Copy the grid and plot the following points: (1, 1), (4, 1), (4, 2), (3, 3), (2, 3), (1, 2).

Join the points in order and label the resulting shape **A**. (2)

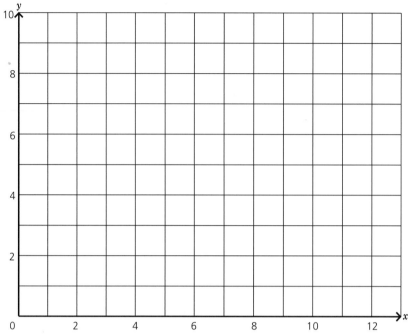

**(ii)** Enlarge shape **A**, using a scale factor of 3, with (1, 0) as the centre of enlargement. Label the enlarged shape **B**. (3)

**(iii)** The area of **A** is 5 cm². What is the area of shape **B**? (3)

22 (i) Copy the grid and draw the triangle with vertices (3, 1), (4, 3) and (1, 3) and label the triangle **A**. (1)

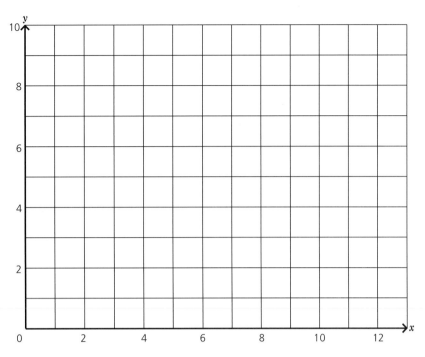

(ii) Enlarge triangle **A** by scale factor 3, using the origin as centre of enlargement, and label the enlarged triangle **B**. (2)

(iii) How many triangles, identical to triangle **A**, could be cut from triangle **B**? (2)

**23** **(i)** On a square grid, copy this diagram and then, with centre *P*, enlarge the shape **S** using a scale factor of 2 (3)

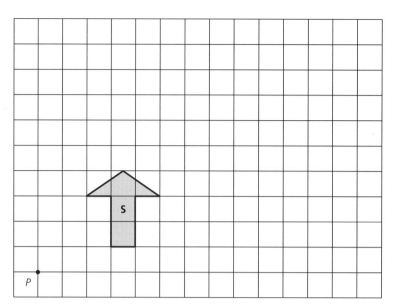

**(ii)** Find the area in squared units of the enlarged shape **T**. (2)

**(iii)** Hence, or otherwise, find the area of the original shape, **S**. (2)

Level 3 ▪ ▦ **24** In the diagram, *ABD* and *BCD* are right-angled triangles.

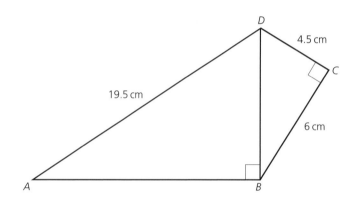

**(i)** Calculate the length of

(a) *BD* (2)

(b) *AB*. (2)

**(ii)** Calculate the area of *ABCD*. (3)

**25** The diagram shows a symmetrical roof frame section made from six lengths of wood, *AE, AC, EC, FB, FC* and *FD*.

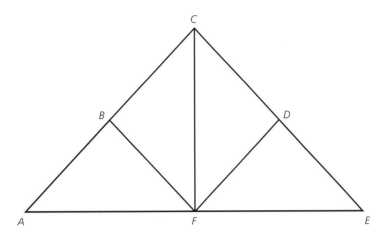

The width, *AE*, of the roof is 8 metres and the height, *FC*, is 3 metres.

Angles *AFC, CBF* and *CDF* are all right angles.

**(i)** Calculate the length of *AC*. (2)

**(ii)** Given that the length of *BC* is 1.8 metres, calculate the length of *BF*. (2)

**(iii)** Calculate the total length of wood required to make the whole roof frame section. (2)

Eight frame sections are placed one metre apart to form the frame for the roof.

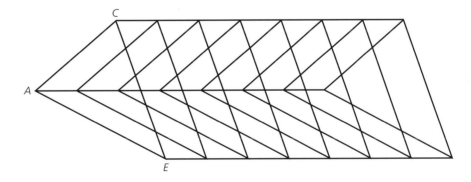

**(iv)** Calculate the volume of the roof space. (3)

**26** The diagonals of a rhombus are 6 cm and 8 cm.

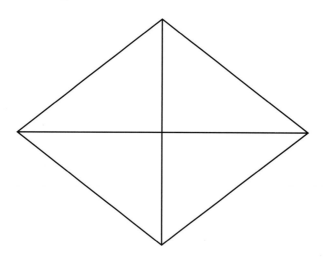

    **(i)** Calculate the length of a side of the rhombus. (2)

    **(ii)** Calculate the area of the rhombus. (2)

## Extension questions

You can try these questions if you have studied **trigonometry** as an extension topic.

### Extension
Questions 27 and 28 are included for interest.

**27** Look at the diagram in question 26. Calculate the angles of the rhombus, giving your answers to the nearest degree. (4)

**28 (a)** A right-angled triangle *ABC* has the dimensions shown.

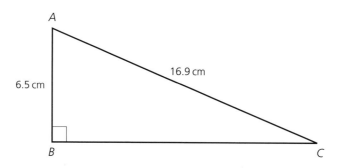

    **(i)** Calculate the length of *BC*. (2)

    **(ii)** Calculate the area of the triangle. (2)

    **(iii)** Calculate angle *ACB*. (3)

**(b)** In triangle *PQR*, angle *PQR* is 48° and angle *PRQ* is 42°.

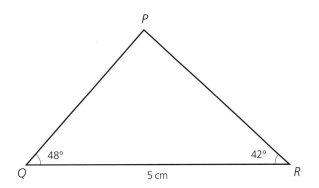

**(i)** Calculate angle *QPR*. (1)

*QR* is 5 cm.

**(ii)** Calculate *PQ*. (2)

**(iii)** Calculate *PR*. (2)

**(iv)** Calculate the area of the triangle. (2)

# Statistics and probability

## 6.1 Statistics

In this section the questions cover the following topics:

- Raw data and tallying
- Bar charts and frequency diagrams
- Range, mean, median and mode
- Pie charts
- Line graphs
- Conversion graphs
- Scatter diagrams

Many questions cover several topics.

*In this section questions should be answered without a calculator except where indicated by* 🖩

1 Louis has the following 20 coins in his savings jar:

| 5p | 20p | 10p | 5p | 20p | 50p | 5p | 10p | 20p | 20p |
|----|-----|-----|----|-----|-----|----|-----|-----|-----|
| 5p | 10p | 50p | 20p | 10p | 5p | 5p | 50p | 50p | 10p |

   (i) Copy and fill in this tally chart. (3)

| Type of coin | Tally | Frequency |
|--------------|-------|-----------|
| 5p | | |
| 10p | | |
| 20p | | |
| 50p | | |
| | Total | |

   (ii) What is the modal value of these coins? (1)

   (iii) What is the median value of these coins? (2)

   (iv) What is the total value, in pence, of all the money? (2)

   (v) What is the mean value of these coins? (2)

2  A group of children counted the number of sweets in each of 30 packets. The results are given in the table. Copy the table.

| 23 | 21 | 19 | 23 | 23 | 21 |
|----|----|----|----|----|----|
| 20 | 23 | 18 | 22 | 21 | 20 |
| 23 | 22 | 19 | 22 | 22 | 22 |
| 22 | 21 | 20 | 19 | 23 | 19 |
| 19 | 20 | 20 | 20 | 22 | 21 |
| **Column totals** | 107 | 107 | 96 | 106 | 111 | |

(i) (a) Add the numbers in the right-hand column. (1)

   (b) The numbers in the first five columns total 527

      What was the total number of sweets? (1)

(ii) What was the mean (average) number of sweets in a packet? (2)

(iii) Copy and complete this tally chart: (3)

| Number of sweets | Tally | Frequency |
|------------------|-------|-----------|
| 18 | | |
| 19 | | |
| 20 | | |
| 21 | | |
| 22 | | |
| 23 | | |

(iv) What is the mode? (1)

3  In a Spanish vocabulary test, the following marks were recorded for the 20 children in the class:

| 3 | 10 | 9 | 9 | 8 | 10 | 4 | 3 | 8 | 5 |
|---|----|----|----|----|----|----|----|----|----|
| 10 | 5 | 8 | 8 | 7 | 10 | 9 | 9 | 6 | 9 |

(i) Copy and complete this frequency table. (4)

| Mark awarded | 1 | 2 | 3 | 4 | 5 | 6 | 7 | 8 | 9 | 10 |
|--------------|---|---|---|---|---|---|---|---|---|----|
| Number of children | | | | | | | | | | |

(ii) State the modal mark. (1)

(iii) Find the median mark. (3)

(iv) Calculate the total number of marks scored by all the children. (2)

(v) Calculate the mean of the marks. (1)

(vi) Why might the class choose to use the mode, rather than the median or mean, when asked about the set of marks? (1)

4  Gina asked the children in Year 8 to choose their favourite classical composer from a list of five composers. She drew a bar chart of the results.

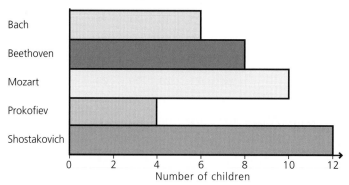

(i)   How many children are there in Year 8? (1)

(ii)  What fraction of the children in Year 8 chose Shostakovich? (2)

(iii) What percentage of the children in Year 8 chose Bach? (2)

(iv)  If the Head selected a child at random from Year 8, what is the probability that the child's favourite composer is either Beethoven or Mozart? (2)

5  The frequency diagram shows the results of a study of the distances travelled by dried peas fired with a pea shooter.

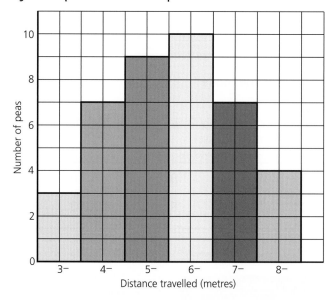

The label '5–' means '5 or more and less than 6'.

(i)   How many peas were fired altogether? (2)

(ii)  What was the shortest possible distance travelled by a pea in this study? (1)

(iii) How many peas travelled eight metres or further? (2)

(iv)  One pea used in the study is chosen at random. What is the probability that it went 8 metres or further? (1)

**6** Jacob's results slip, given here, shows the percentage marks on his end-of-term examination papers.

| Jacob | | | |
|---|---|---|---|
| English 1 | 62 | French | 59 |
| English 2 | 55 | Spanish | 64 |
| Mathematics 1 | 83 | History | 58 |
| Mathematics 2 | 73 | Geography | 41 |
| Science | 58 | Religious Studies | 60 |

**(i)** **(a)** What was Jacob's total mark? (2)

**(b)** What was his mean (average) mark? (2)

Diana took the same papers. Her mean mark was 65.8%.

**(ii)** **(a)** What was Diana's total mark for the ten papers? (1)

Diana spilled ink on her results slip, as shown here.

| Diana | | | |
|---|---|---|---|
| English 1 | 59 | French | |
| English 2 | 57 | Spanish | |
| Mathematics 1 | 75 | History | 58 |
| Mathematics 2 | 68 | Geography | 46 |
| Science | 63 | Religious Studies | 63 |

Diana cannot remember what her French and Spanish marks were but she can remember that the French mark was one mark higher than the Spanish mark.

**(b)** Calculate Diana's French mark. (3)

**(iii)** In which papers was Diana's mark higher than Jacob's? (3)

**7** The mean mass of ten children is 34.4 kg.

**(i)** What is the total mass of the children? (2)

When two of the children leave, the mean mass of the eight remaining children falls to 30 kg.

**(ii)** What is the total mass of the eight remaining children? (1)

**(iii)** What is the mean mass of the two children who left the class? (2)

8   Twelve girls ran 100 metres and their times, in seconds, were recorded.

| 14.6 | 15.2 | 15.3 | 14.1 | 16.9 | 16.2 |
| 15.6 | 15.5 | 14.7 | 14.9 | 17.6 | 17.9 |

(i)   For this set of data, calculate

(a) the range of times                                                                    (1)

(b) the median time                                                                       (2)

(c) the mean time.                                                                        (2)

(ii)  (a) Calculate the average speed of the fastest girl in the group.

       Give your answer in metres per second.                                            (2)

(b) Convert your answer to part (ii) (a) into kilometres per hour.      (2)

9   On a farm, 65% of the animals are sheep, 30% are cows and the rest are pigs.

(i)   What percentage of the animals on the farm are pigs?                          (1)

(ii)  The farmer wants to draw a pie chart to represent his livestock. What
      sized angle would represent

(a) 10%                                                                                    (1)

(b) 5% of the animals?                                                                     (1)

(iii) Starting with a copy of this diagram, draw a pie chart, clearly marking
      the angles and the sectors representing each type of farm animal.       (3)

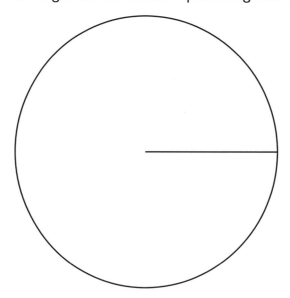

10  A total of 72 children took part in an athletics competition. Each child was entered for only one event. A pie chart is drawn to show the numbers of children in the different events.

(i)   How many degrees would represent each child in the pie chart?   (2)

Seven children entered the long jump.

(ii)  What is the angle of the sector representing these children?   (2)

The children who entered the 100 metres are represented by a sector of angle 55°.

(iii) How many children entered the 100 metres?   (2)

11  An art book has 120 pages. Tom counted the number of pictures on each page. This pie chart shows that 45% of the pages have one picture, 20% of the pages have two pictures, and so on.

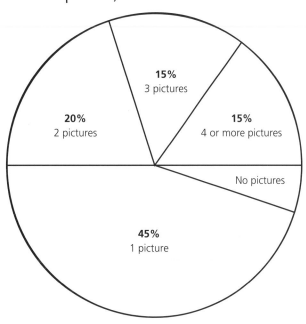

(i)   Find the percentage of the pages with no pictures.   (2)

(ii)  How many pages have one picture?   (2)

(iii) Calculate the size of the angle of the sector which represents the pages with two pictures.   (3)

**12** This pie chart shows the percentage of the 24 hours of a typical day spent by Jackie's dog, Rosie, in a variety of activities.

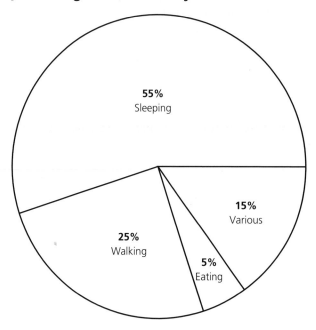

- **(i)** How many hours are represented by the 'walking' sector? (2)

- **(ii)** How much time is spent on 'sleeping'?

  Give your answer in hours and minutes. (2)

- **(iii)** What angle does the 'various' sector make at the centre? (2)

**13** 180 girls were asked to name their favourite sport. The choices were swimming, hockey, tennis and netball. The results are to be shown in a pie chart, part of which has been drawn.

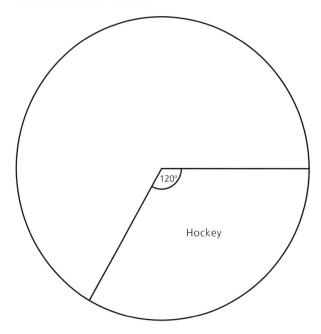

- **(i)** How many girls chose hockey? (2)

A quarter of the girls chose netball.

**(ii)** Copy the diagram, taking care that the angle of the sector representing Hockey is exactly 120°, and then draw a sector to represent Netball, marking it clearly with both the angle and the sport. (2)

Of the remainder, twice as many chose swimming as chose tennis.

**(iii)** Show these sectors in the pie chart, marking them clearly with both the angle and the sport. (4)

14 On a ramble, 72 people each chose a sandwich. They had the choice of ham, cheese, tuna, salad or chicken. Part of a pie chart showing their choice of sandwich is shown here.

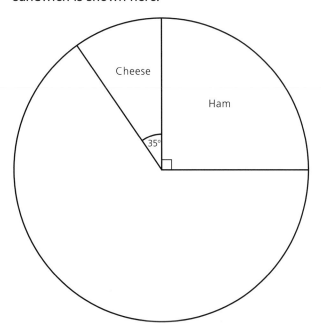

**(i)** How many people chose a ham sandwich? (1)

Twice as many people chose tuna as chose cheese.

Eleven chose a salad sandwich.

The remainder chose a chicken sandwich.

**(ii)** Copy and complete the pie chart, writing the name of the sandwich filling and the size of the angle in each sector. (4)

**15** Each week, Harriet puts a third of her pocket money towards clothes, a quarter towards DVDs for her collection, and the rest she puts in the bank.

(i) Starting with a copy of this diagram, draw a pie chart to display the data, marking clearly the angle of each sector and labelling each of the sectors. (3)

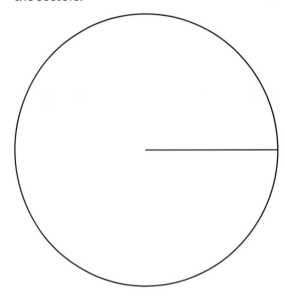

Harriet receives £15.00 pocket money per week.

(ii) How much does Harriet bank each week? (2)

When Harriet's pocket money is raised by 20% per week, she splits the increase equally between her clothes fund and her bank savings.

(iii) How much does she now bank each week? (3)

**16** The temperature was measured at noon on each of the first ten days in March and the measurements are recorded in this table.

| Date | 1st | 2nd | 3rd | 4th | 5th | 6th | 7th | 8th | 9th | 10th |
|---|---|---|---|---|---|---|---|---|---|---|
| Temperature (°C) | 9 | 7 | 7 | 6 | 7 | 8 | 9 | 7 | 10 | 9 |

(i) What is

(a) the range of temperatures (2)

(b) the modal temperature? (1)

(ii) Calculate

(a) the median temperature (2)

(b) the mean temperature. (2)

(iii) Copy and complete this line graph. (2)

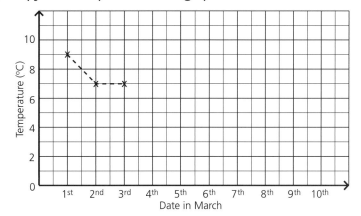

17 This conversion graph can be used to change centimetres to inches.

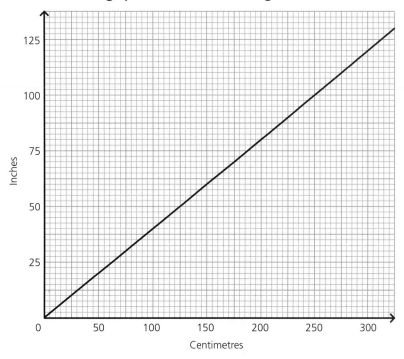

(i) Copy and complete the following statement.

The graph passes through the reference points (0, _ ) and (250, _ ) (1)

(ii) Copy the graph and then use it to answer the following questions, showing clearly where you take your readings.

(a) How many inches are equivalent to 260 centimetres? (2)

(b) There are 12 inches in 1 foot. Simon's height is 5 feet 10 inches.

What is his height measured in centimetres? (3)

**18** In the land of Glum, 4 yuks are equivalent to 11 ughs.

   **(i)**   Calculate how many ughs are equivalent to 20 yuks.         (2)

   **(ii)**   On graph paper, copy this diagram, complete the labelling of the axes and then draw a graph to convert yuks into ughs.         (3)

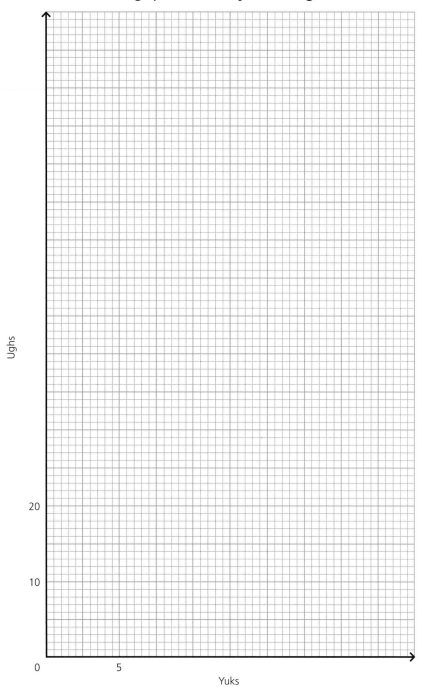

   **(iii)** From your graph, marking clearly where you take your readings, find

      **(a)** the value in ughs of 11 yuks         (1)

      **(b)** the value in yuks of 45 ughs.         (1)

**19** Oaty biscuits come in 250 gram packets. The label on a packet of Oaty biscuits includes the following nutrition information.

| Energy (calories) | per 100 g | per biscuit |
|---|---|---|
| | 480 | 60 |

(i) What is the mass of one Oaty biscuit? (2)

Johnny has started making this table, so that he can draw graphs relating the number of biscuits eaten to mass and to energy.

| | 1 biscuit | 100 g | 250 g packet |
|---|---|---|---|
| Number of biscuits | 1 | | |
| Mass (g) | | 100 | 250 |
| Energy (calories) | 60 | | |

(ii) Copy and complete Johnny's table. (5)

**20** Brian's car uses on average 8 litres of petrol every 100 kilometres.

(i) Using this fact, calculate the number of litres of petrol that Brian's car uses to travel 500 kilometres. (2)

(ii) Copy this grid and then draw a line graph to show how much petrol Brian's car uses for distances up to 600 km. (2)

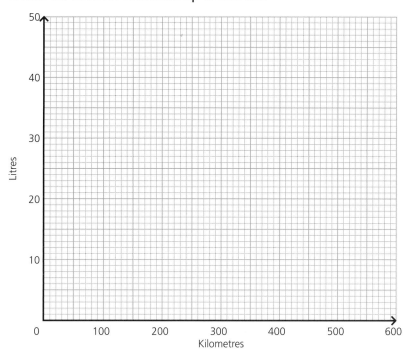

(iii) Use your graph to answer the following, showing clearly where you take your readings.

(a) How far will the car travel on 36 litres of petrol? (2)

(b) Brian wants to travel 335 kilometres. His car has 17 litres of petrol in the tank. How much more petrol will he need? (2)

**21** This scatter graph shows the marks for English examination papers 1 and 2 of a class of 20 children.

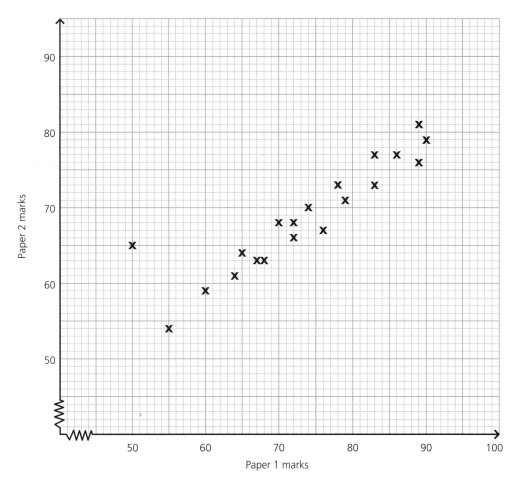

(i)    What type of correlation does this scatter graph suggest?                    (1)

(ii)   How many pupils scored a higher mark on Paper 2 than on Paper 1?    (1)

(iii)  Place your ruler on the scatter graph so that the ruler edge lies along the line of best fit and then copy and complete the following statement.

The line of best fit passes through (or very close to) the points (55, _ ) and (90, _ ).                                                                 (2)

(iv)   Peter was absent for Paper 2 but he scored 73 marks on Paper 1.

Use the scatter graph to estimate the mark Peter might have obtained on Paper 2 if he was typical of this group.                           (2)

**22** The petrol gauge in Mr Hillard's new car indicates the number of litres of petrol in the tank. On a car journey from Inverness to Norwich, Mr Hillard recorded the petrol gauge reading at different points on the journey. This table gives the results.

| Distance travelled (miles) | Fuel left (litres) |
|---|---|
| 0 | 50 |
| 120 | 31 |
| 215 | 22 |
| 295 | 10 |
| 340 | 5 |

**(i)** Copy this grid and plot the results. (3)

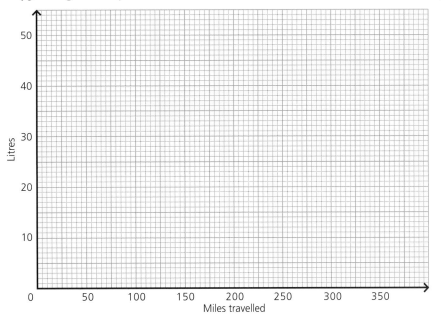

**(ii)** On your grid draw the line of best fit. (1)

**(iii)** Use your graph, showing clearly where you take your readings, to estimate how many litres of petrol he had used when he had travelled 250 miles. (2)

**(iv)** By extending your line of best fit, estimate the distance that would be travelled by Mr Hillard's car before it ran out of fuel, if he did not stop for petrol. (2)

**23** This scatter diagram shows the mean ages and scores of the teams in a general knowledge quiz.

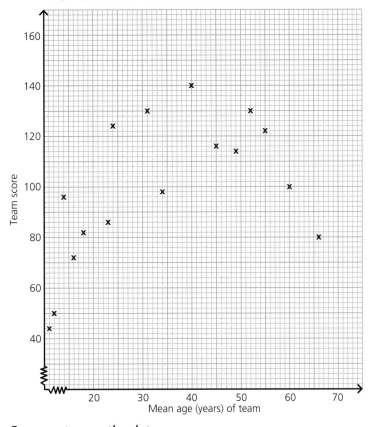

Comment upon the data. (3)

**24** A group of 14 friends are to attend a function where they will wear Scottish dress including kilt, jacket and brogues (shoes). They send the following data to a kilt hire shop.

| Name | Waist (inches) | Chest (inches) | Shoe size |
|------|----------------|----------------|-----------|
| Alan | 34 | 38 | 9 |
| Blair | 38 | 42 | $9\frac{1}{2}$ |
| Calum | 32 | 38 | 8 |
| Don | 36 | 44 | 10 |
| Ewan | 42 | 50 | $10\frac{1}{2}$ |
| Fred | 34 | 40 | 10 |
| George | 46 | 52 | 12 |
| Hamish | 32 | 40 | 9 |
| Iain | 36 | 38 | 9 |
| James | 40 | 40 | 10 |
| Ken | 34 | 38 | $8\frac{1}{2}$ |
| Liam | 38 | 44 | 9 |
| Matt | 44 | 42 | $9\frac{1}{2}$ |
| Nat | 40 | 44 | 10 |

**(i)** From the data in the table, find

    **(a)** the modal waist size (1)

    **(b)** the range of chest sizes (1)

    **(c)** the median shoe size. (2)

**(ii)** **(a)** Choosing suitable scales for the axes, draw a scatter diagram showing waist measurements and shoe sizes. (5)

    **(b)** Draw a line of best fit on your scatter diagram. (1)

    **(c)** Comment upon the correlation. (2)

**(iii)** When the group arrives at the hire shop, Fred picks up a pair of shoes at random. What is the probability that he picks up a pair of the correct size? (2)

# 6.2 Probability

In this section the questions cover the following topics:

- Outcomes of events
- Probability

**1** Sally has four gloves in her pocket, one blue pair and a similar green pair, which have been jumbled together.

    **(i)** If Sally pulls out two gloves at random, she might pull out a left blue and a right green. Write down the other five different combinations of gloves which Sally might select. (3)

**(ii)** What is the probability that Sally pulls out

    **(a)** the pair of blue gloves     (1)

    **(b)** a pair of gloves of the same colour?     (1)

**2** Mr Styk is choosing two players to fill the positions of goalkeeper and sweeper in the hockey team. Four boys are keen to be chosen.

Peter (P) could only play goalkeeper, but Quentin (Q), Rory (R) and Stuart (S) could play in either position.

    **(i)** List the different ways in which the hockey team could be completed. Put the two possibilities which are done for you into your list.     (3)

| Goalkeeper | Sweeper |
|---|---|
| P | Q |
| | |
| | |
| Q | R |
| | |
| | |
| | |
| | |
| | |

Each of the different ways in which the team could be completed is put in the goalkeeper's helmet and Mr Styk picks one at random.

    **(ii)** What is the probability that Rory is chosen to play?     (2)

    **(iii)** If Rory is chosen to play, what is the probability that Stuart is *not* chosen to play?     (2)

**3 (a)** A bag contains three coins, 1p, 2p and 5p. A second bag contains just two coins, 1p and 5p. One coin is taken at random from each bag.

    **(i)** Copy and complete the diagram to show all the possible pairs of coins which could be taken.     (1)

| Second bag | First bag | | |
|---|---|---|---|
| | 1p | 2p | 5p |
| 1p | 1p, 1p | 2p, 1p | |
| 5p | 1p, 5p | | 5p, 5p |

    **(ii)** How many pairs of coins contain only one 1p?     (1)

    **(iii)** What is the probability that just one of the two coins chosen is a 1p?     (1)

    **(iv)** What is the probability that the value of the coins chosen totals 6p?     (1)

(v) What is the probability that a total amount of more than 4p is chosen? (1)

(b) A number is chosen at random from all the numbers from 20 to 49 inclusive.

    (i) What is the probability that the number is greater than 43? (2)

    (ii) What is the probability that the number is prime? (2)

4 Copy the probability scale and:

    (i) mark with the letter A the probability of getting heads when a fair fifty-pence coin is tossed (1)

0              $\frac{1}{2}$              1

    (ii) mark with the letter B the probability of rolling a 2 with an ordinary fair die (1)

    (iii) mark with the letter C the probability of getting an odd product when two ordinary fair dice are rolled together and their scores multiplied. (4)

5 (a) Sarah organises a fundraising raffle for a teddy bear. She sells 18 tickets altogether. Philip buys 5 tickets and Roxanne buys 1 ticket.

    (i) What is the probability that Philip wins the bear? (1)

    (ii) What is the probability that neither Philip nor Roxanne wins the bear? (2)

(b) Connor is playing a game in which he tosses a fair coin 4 times. He wins if he gets 'tails' on each of the 4 tosses.

    (i) What is the probability that he gets 'tails' on his first toss? (1)

    (ii) Connor has tossed 3 'tails' in a row. Which of these statements best describes his chance of winning the game with the last toss? (1)

**very unlikely    unlikely    even chance    likely    very likely**

6 A number is chosen at random from the integers 21 to 40 inclusive.

    (i) What is the probability that the number is prime? (2)

    (ii) What is the probability that the number is an even multiple of 3? (2)

7 The letters of the word 'calculator' are written on cards as shown.

| C | A | L | C | U | L | A | T | O | R |

The cards are shuffled and Clare takes one card at random.

(i) What is the probability that Clare takes a letter A? (1)

The card is replaced and Clare takes another card, again at random.

(ii) What is the probability that this letter has at least one line of symmetry? (2)

Again, the card is replaced and Clare takes another card. This time she takes a letter R and she *does not* replace the card. Clare takes a second card, so that she will have two cards in her hand.

(iii) What is the probability that both letters will be in her name? (2)

8 On a snooker table there are 15 red balls and one each of white, yellow, green, brown, blue, pink and black.

James, who is blindfolded, picks one ball at random from the table.

(i) What is the probability that the ball is

(a) white (1)

(b) not red? (1)

James picked the blue ball and put it in his pocket.

Faye now removes two thirds of the red balls, the pink ball and the black ball.

James then picks another ball at random.

(ii) What is the probability that he picks the white ball? (2)

9 In this Venn diagram:

$\mathscr{E}$ = {integers from 30 to 40 inclusive}

A = {even numbers}

B = {multiples of 3}

C = {multiples of 5}

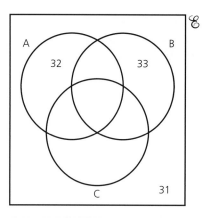

(i) List all the members of

(a) set A (1)

(b) set B (1)

(c) set C (1)

(ii) Copy and complete the Venn diagram. (3)

(iii) A number is chosen at random from ℰ.

What is the probability that the number chosen is a multiple of both 2 and 3? (2)

10 Of the 18 boys in class 6A

● 4 play in the school soccer team but not in the school rugby team

● 2 do not play in either team

● 11 play in one team only.

(i) Represent this information in

(a) a Carroll diagram (2)

(b) a Venn diagram (2)

(ii) How many boys in class 6A

(a) play in both teams (1)

(b) play in the rugby team (1)

(c) do not play in the soccer team? (1)

(iii) A boy in 6A is chosen at random. What is the probability that he plays in the rugby team? (2)